Transition Metal Oxides at Low Dimensions

This book covers enhanced properties and cutting-edge applications of transition metal oxides (TMOs) at low dimensions (2D, 1D, and 0D). The electronic bandgap modification and emergence of novel and fascinating properties due to reduced dimension are discussed in detail. It covers various synthesis procedures and advanced technical tools for the characterization of the TMOs at lower dimensions and enhanced properties of the TMOs at lower dimensions realizing ground-breaking applications in advanced energy storage and energy harvesting, catalysis, and chemical sensing.

Features:

- Covers applications and possible devices made up of nanoclusters/quantum dots (QDs) and 1D and 2D TMOs.
- Discusses low-dimensional TMOs for advanced properties.
- Explores next-generation advancement in green energy, gas/chemical sensing, and photocatalysis.
- Reviews low-cost TMO-based devices with advantage over electronic-grade transition metal dichalcogenides (TMDs).
- Details physics of 1D TMOs including perspectives from spin wave and charge density wave.

This book is aimed at researchers and graduate students in materials science.

Transition Metal Oxides at Low Dimensions

Augmented Properties and Advanced Applications

Raktima Basu and Sandip Dhara

CRC Press
Taylor & Francis Group
Boca Raton London New York

CRC Press is an imprint of the
Taylor & Francis Group, an **informa** business

Designed Cover Image: Shutterstock

First edition published 2025
by CRC Press
2385 NW Executive Center Drive, Suite 320, Boca Raton FL 33431

and by CRC Press
4 Park Square, Milton Park, Abingdon, Oxon, OX14 4RN

CRC Press is an imprint of Taylor & Francis Group, LLC

ISBN: 9781032796611 (hbk)
ISBN: 9781041034018 (pbk)
ISBN: 9781003623694 (ebk)

DOI: 10.1201/9781003623694

Typeset in Times
by Newgen Publishing UK

Contents

Preface....vii
Acknowledgments....ix
About the Authors....xi
Introduction....xiii

Chapter 1 A Perspective on Transition Metal Oxides at Reduced Dimensions....1

1.1 Introduction....1
1.2 Surface Properties....2
1.3 Electronic Properties....2
1.4 Magnetic Properties....3
1.5 Optical Properties....4
1.6 Applications....4

Chapter 2 Transition Metal Oxides at 2D: Structure, Properties, and Applications....7

2.1 Introduction....7
2.2 Van der Waals-Bonded 2D Transition Metal Oxides....8
2.3 Engineered 2D Transition Metal Oxides....21

Chapter 3 Transition Metal Oxides at 1D....34

3.1 Quasi-1D TMOs: Charge Density Wave....34
3.2 Energy Applications....40
3.3 Gas Sensing....44

Chapter 4 Transition Metal Oxides at 0D....55

4.1 TMO Quantum Dots....55
4.2 TMO Nanoclusters....61

Chapter 5 Heterostructures with Transition Metal Oxides....71

5.1 Introduction....71
5.2 TMD/TMO Heterostructures....72
5.3 Graphene/2D TMO Heterostructures....74
5.4 MXene/2D TMO Heterostructures....80

Chapter 6 Perovskite Oxides: Properties at 1D and 2D 90

6.1 Introduction 90
6.2 1D Perovskite Oxides 90
6.3 2D Perovskite Oxides 95

Chapter 7 Few Special Transition Metal Oxides: 1D Spin Dynamics 101

7.1 VO_2: 1D Finite-Sized Heisenberg Spin Chain 101
7.2 $CuGeO_3$: 1D Linear Spin-Peierls Dimers 108

Chapter 8 Conclusion and Future Direction 117

Index 121

Preface

Transition metal oxides (TMOs) have played a vital role in the progress of materials science, providing technological advances and developing an understanding of various new physical phenomena in condensed matter physics.

The book covers the properties and advanced applications of TMOs at reduced dimensions. Modification of electronic band structures at low dimensions is conferred for the expected changes in their magnetic, electrical, and optical properties. Most TMOs are associated with one or more phase transitions, that is, metal-to-insulator transition (MIT), structural phase transition (SPT), and magnetic and optical transitions, as a function of temperature, pressure, electric/magnetic field, and so on. The correlation of the critical parameters of the transitions with the reduced dimension is thoroughly discussed. The reduction in dimension not only discloses novel and fascinating physical properties but also opens new doors to modern future technologies that rely on the size of a material. The book depicts the underlying physics behind the enhanced properties of the TMOs at 2D, 1D, and 0D, realizing groundbreaking applications in advanced energy storage and energy harvesting, catalysis, chemical sensing, and so on. The book emphasizes the TMO heterostructures, especially focusing on their potential applications due to interface-induced phenomena and reduced dimensionality. Perovskite TMOs are promising candidates for various photonic and optoelectronics applications including solar cells. The perovskite TMOs prepared at the nanoscale have boosted efficiency and novel properties. The book also covers a few unique TMOs in which the reduction in dimension alters the system's spin dynamics and local band structure, resulting in new phase transitions. The aim of the book is to stimulate the search for low-dimensional TMOs with unique and customizable physical properties and cutting-edge applications.

Acknowledgments

We acknowledge our family members for accommodating us at various stages of writing the book. We appreciate their co-operation. We also thank Dr. Reshma P. R., Research Associate, Materials Science Group, IGCAR, for her suggestions in the early stage of determining the contents and title of the book.

About the Authors

Raktima Basu is a physicist in the field of experimental condensed matter with a specialization in phase transition in correlated metal oxide nanostructures at extreme conditions (pressure and temperature) and their applications. At present, she is working as an assistant professor of physics at Magadh University, Bodhgaya, Bihar. She completed her M.Sc. in physics in 2014 from the University of Burdwan (ranked 2nd) and achieved her Ph.D. in 2020 from Homi Bhabha National Institute (HBNI). She was awarded the Outstanding Doctoral Student Award (2021) from HBNI, the Foundation Day 'Best Thesis' award of Indira Gandhi Centre for Atomic Research (IGCAR), and the 'Young Scientist Award' (2018) from Dr. K. V. Rao Scientific Society, India. She was also awarded the National Post-Doctorate Fellowship, and she worked at the Department of Physical Sciences, Indian Institute of Science Education and Research Kolkata (2021–2023), India. She also worked as a Research Associate II at the Surface Physics and Material Science Division, Saha Institute of Nuclear Physics, Kolkata, India (2023–2024). She received five best presentation awards at national and international conferences. So far, she has more than 27 publications in peer-reviewed international journals and conference proceedings.

Sandip Dhara completed his Ph.D. in magneto-optic recording materials from the National Physical Laboratory, New Delhi, in 1994 and joined the Indira Gandhi Centre for Atomic Research (IGCAR), Kalpakkam, in 1996. He is the Director of the Materials Science Group, IGCAR, and a Senior Professor at Homi Bhabha National Institute. He was an Associate Professor at National Cheng Kung University, Taiwan (2007–2008). He was also a Visiting Professor at the University of Valencia, Spain; Technical University of Chemnitz, Germany; and the Institute of Atomic and Molecular Sciences, Academia Sinica, Taiwan. Dr. Dhara has more than 300 research publications, including several patents, review articles, books, book chapters, and conference proceedings. He has delivered more than 100 invited and plenary talks and is the recipient of the Homi Bhabha Gold Medal Award 2019 and the Department of Atomic Energy-Scientific Research Council (DAE-SRC) Outstanding Researcher Award for 2012. He is also on the editorial board of many national and international scientific journals. Dr. Dhara has specialized in studies of light-matter interaction at the nanoscale in the sub-diffraction limit. He is a Fellow of the Institute of Physics (FInstP), UK, and a Fellow of the Royal Society of Chemistry, London. He is also a Senior Member of Optica (old Optical Society of America).

Introduction

Transition metal oxides (TMOs) have played a crucial role in the long saga of materials science by providing technological advances leading to their immense utility in daily life or by developing a deep understanding of various aspects of condensed matter physics. TMOs are functional materials and play an important role in the progress of modern civilization. At the same time, the unique nature of outermost *d*-electrons in TMOs makes them materials of immense interest with exceptional electronic, magnetic, and optical properties depending on their crystalline structure. Moreover, owing to strong electronic correlation and structure-property coupling, TMOs have become the test bed of various cutting-edge physical concepts of high-temperature superconductivity, spin wave, charge density wave, and colossal magnetoresistance, among others. We conceived the idea of writing the book by looking into the various important aspects of material advancement and physical concepts involving TMOs at their reduced dimension.

We start by highlighting various important physical properties of TMOs at low dimensions. In this regard, we discuss their surface properties and manifestation with the variation of charge polarity, degree of coordinative unsaturation, and defect sites present. We also elaborate on the electronic and magnetic properties of TMOs at reduced dimensions using the Heisenberg model and the need for Mott-Hubbard approximation. Bandgap tailoring and unusual optical absorption in various TMOs at low dimensions are also considered.

Coming down from the bulk properties, naturally occurring two-dimensional (2D) TMOs are considered first, describing their structural evolution, optical and electrical properties of van der Waals (vdW)-bonded and engineered materials, and applications, including industrially important catalysis. One has to keep in mind that the growth of monolayer TMO is more complicated than that reported in the case of transition metal dichalcogenides (TMDs) because of the ionic bonding in TMO, leading to the creation of a dangling bond. Thus, the numbers of vdW-bonded and engineered 2D TMOs are limited and are still in an exploratory stage, facing various challenges, including the growth process and their useful applications. These 2D TMOs, however, showed noble optical and electrical properties and enhanced photocatalytic activity.

We elaborate on charge density wave (CDW), a relatively new concept of physics in the quasi-1D regime of TMOs. Electron-phonon interactions leading to CDW are considered for 1D metallic blue bronze and in quasi-1D stripe domains of finite width involving dynamical modulation of spin and charge in hole-doped superconducting cuprates and Ruddelsden–Popper-type layered structure. A complete understanding of CDW explaining diverse physical phenomena has yet to be developed. In the context of 1D TMOs, we also discuss electrochemical and photovoltaic energy storage applications along with catalytic activities. Studies related to gas sensing in the resistive mode with the advantage of unidirectional electrical conduction in the 1D structures are also described for various TMOs.

Quantum dots (QDs), which are ideal 0D material, have attracted applauding attention in the recent past, with the Nobel Prize based on them being announced most recently in 2023. Thus, TMO nanostructures at 0D must grab attention. These 0D nanostructures with significantly high effective areas showed unsaturated chemical bonds at the surface, lowering activation energy suitable for catalytic and sensing applications. Photocatalytic and other energy applications of various 0D TMOs along with the chemiresistive gas-sensing applications are elaborated. QDs of TMOs draw special attention due to their inherent oxygen non-stoichiometry, which promotes chemical activity, relatively high carrier lifetime, and stability. The well-celebrated optical confinement in QDs and nanoclusters is potentially useful for photocatalytic and energy applications.

Semiconductor heterostructures artificially made of two or more dissimilar materials create novel interface states that are often different from those observed in thermodynamically stable bulk materials of individual constituents. The 2D heterostructures of TMOs with TMDs, graphene, metal carbide, nitrides, and carbonitrides (MXenes) are discussed. A novel phenomenon of modulation doping of 2D-TMD/TMO heterostructure exhibiting charge carrier injection and carrier transport in the defect-free 2D layer is narrated. Advanced energy storage applications in 2D graphene/TMO heterostructures are also deliberated. A novel capacitive deionization used in desalination and supercapacitive energy storage using MXene/TMO heterostructures are also discussed.

We pay special attention to the recent developments in perovskite oxides and their potential applications as multiferroics. In this book, we cover the growth process and novel physical properties of 1D and 2D perovskite oxides. In the 1D form, nanowires and nanotubes of perovskite oxides containing transition metals are considered. However, a complete understanding of structure-property correlation is in its initial state as the growth mechanism involved in the evolution of the 1D ternary phase is very complex. On the other hand, the growth of 2D transition metal containing perovskite oxides is well understood, and the formation of monolayer is described. Easy transfer of the reported monolayers to suitable substrates made it more attractive for the device applications.

A few special TMOs having 1D Heisenberg spin ½ chains, namely VO_2 and $CuGeO_3$, are also considered. The 1D spin ½ chains lead to spin-Peierls type structural phase transition (SPT). A metal-to-insulator transition (MIT) and SPT are reported simultaneously in VO_2, with a chicken and egg debate being deliberated on which takes control over the phase transition. A collective spinon vibration is recorded for both materials. Pressure-induced phase transition in $CuGeO_3$ is also elaborated. Applications of these special TMOs are also described. The final chapter offers a summary and the expected future developments.

When one says technological advances with TMOs and their utilization, it primarily means providing material comfort, safety, and security to living beings. We can also emphasize here that TMOs are abundantly present in the earth's crust as a boon for living beings. Dimensionality plays a role in switching to modulate various important physical properties, namely, electronic, magnetic, and optical properties. Thus, an effort is made here to generate overall knowledge for various TMO-nanostructured materials in developing next-generation devices.

1 A Perspective on Transition Metal Oxides at Reduced Dimensions

1.1 INTRODUCTION

Transition metals belong to the *d*-block series of the periodic table with partially filled *d*-shell (e.g., Fe, Co, Ni, Ti, Zr, V, W, Mo, etc.). Oxides of transition metals attracted tremendous attention due to their extraordinary chemical and physical properties, mostly because of the unique nature of outermost *d*-electrons. The metal-ligand bonds, depending on the corresponding oxidation state, can control the nature of the electrical character (ionic to metallic).[1] Therefore, the electric properties of transition metal oxides (TMOs) strongly depend on their structural properties. Moreover, TMOs also display outstanding optical, magnetic, and electronic properties depending on the phase in which they are stabilized. The variation in crystal structure via external parameters, for example, temperature, pressure, chemical composition, or strain, plays a crucial role to induce various phase transitions. Due to strong electronic correlation and structure-property coupling, TMOs find application in various new-age fields, including high-temperature superconductivity, multiferroics, colossal magnetoresistance, ferro/piezoelectricity, microelectronics, energy storage, sensors, and so on.[2,3] As the physical and chemical properties of TMOs depend on the size and morphology, engineering TMOs in lower dimension not only retains their basic characteristics but also induces novel effects due to nano-dimension.[4,5] A perovskite TMO was proposed ~50 years ago to exhibit exceptional electrical conductivity and excellent catalytic properties toward alkaline water.[6] The reduction in dimension introduces novel properties in TMOs by enhancing the surface area, decreasing electron-hole recombination, increasing conductivities, and others.[7-10] As a result, in the recent years, TMO-based nanostructures have gained significant attention for their use in multipurpose functional applications. Many TMOs in zero (0D), one- (1D), and two-dimensional (2D) structures are reported in the literature for applications such as electrode material for supercapacitors,[11] antibacterial agents,[12] sensors,[13] photocatalytic degradation,[14] photovoltaic cells,[15] and many others.[16]

In this book, we discuss important physical and chemical properties of TMOs especially at lower dimensions along with their applications.

DOI: 10.1201/9781003623694-1

1.2 SURFACE PROPERTIES

The surface structure of TMOs, especially at reduced dimensions (nanoscale regime), is essential due to the interactions of the surfaces. The surfaces are generated by cleavages, especially along a particular plane and, therefore, may differ from the bulk structure.[17] For example, if a binary TMO is cleaved to generate two new surfaces, the structure of the two newly created surfaces may or may not be the same, though the charge remains neutral. The surfaces may be polar or non-polar depending on the presence and absence of dipoles at the surfaces, respectively.[18] Rocksalt (100), the rutile (100), (110), (001), and perovskite (100) surfaces are non-polar. On the other hand, the rocksalt (111) surface is an example of a polar surface. However, a non-polar surface is more stable than a polar surface as dipole moment increases the surface Gibbs energy. Moreover, oxygen polar surfaces are more stable than metal polar surfaces due to more polarizability of oxygen. Along with polarity, surface properties depend also on the degree of coordinative unsaturation and defect sites.[19] The degree of coordinative unsaturation of a surface depends on the number of bonds broken to form the surface. Therefore, the stabilization of the cation and the overall stability of the surface rely on the degree of coordinative unsaturation. For example, a lower degree of coordinative unsaturation makes the rutile (110) surface more stable than the rutile (100) and (001) surfaces. Defect sites also affect the stability of TMO surfaces. Point defect sites, especially in rocksalt-type surfaces, are more prone to interfere than those of steps and kinks-type defects. However, in rutile surfaces, the most common type of defect is oxygen vacancies. Oxygen vacancies may cause a reduction in surface cations, leading to variations in electronic energy levels.[20]

1.3 ELECTRONIC PROPERTIES

The electronic properties of TMOs mostly depend on the narrow electronic bands formed by the small overlap between the metal *d* and the oxygen *p* levels. Depending on the overlapped ligand 2*p* and metal *d* bands, TMOs may be good metals (ReO_3), semiconductors (CuO), insulators (TiO_2), and even superconductors ($YBa_2Cu_3O_7$). Few special TMOs also exhibit a metal-to-insulator transition (MIT) as a function of temperature (VO_2), pressure (V_2O_3), or composition (Na_xWO_3). Moreover, in the quasi-1D and 2D limit, the appearance of charge density wave, an exotic quantum phenomenon, attracted a great deal of attention in superconducting cuprate and other Ruddelsden–Popper-type layered oxide structures.[21,22] The electronic structure of such materials is generally explained either by band theory (collective electron system) or by the crystal (ligand)-field theory (localized electron system).[23] According to band theory, the Coulomb repulsion energy (U) required to transfer a valence electron from one orbital to another singly occupied orbital on an equivalent site is small compared with the bandwidth (W), as in the case of outer *s* and *p* electrons. On the other hand, the ligand field theory is applicable for outer *f* electrons, where U is large compared with W. However, for outer *d* electrons, which exhibit both itinerant as well as localized electron properties, U is of the order of W, and we have strongly correlated electrons in solids. The strong electronic correlation in the narrow electronic bands distinguishes the TMOs from normal metals, alloys, covalent semiconductors, and ionic insulators. Moreover, the polarizability of oxygen (O^{1-} and O^{2-}) also plays an important role and may induce polaronic or bipolaronic effects, thereby controlling the mobility and strength of correlation. In reduced dimension, the only states near the Fermi energy thus are *d*-electron states. Bloch-Wilson theory of conductivity, which explains the conductivity of most materials very well, however, fails to describe the insulating nature of many TMOs with partially filled *d* bands.

A plausible explanation was provided by Mott (in 1961),[24] who pointed out that in narrow-bandwidth TMOs, the electrons remain localized around particular ion cores without being itinerant. Thus, most TMOs, which show insulating behaviour even with partially filled *d* states, are called Mott insulators. The Mott insulators with extremely narrow *d* bands should include electronic correlation in energy-band structure calculation. An appropriate band calculation model for TMOs with effects of electronic correlations was proposed by Hubbard with approximate Hamiltonian (Eqn. 1.1):

$$H = -t\sum_{i,\sigma}\left(\hat{c}^{\dagger}_{i,\sigma}\,\hat{c}_{i+1,\sigma} + \hat{c}^{\dagger}_{i+1,\sigma}\,\hat{c}_{i,\sigma}\right) + U\sum_{i}\hat{n}_{i\uparrow}\,\hat{n}_{i\downarrow} \tag{1.1}$$

where $\hat{c}^{\dagger}_{i,\sigma}\hat{c}_{i,\sigma}$ is the spin-density operator for spin σ on the *i*-th site, $\hat{n}_i$ is the density operator, and *t* is the hopping integral. For a small value of *W/U*, the band splits into two narrow sub-bands, separated by an energy ~*U*, and the materials behave as an insulator, namely the Mott insulator. However, as *W/U* increases to $2/\sqrt{3}$, the narrow sub-bands overlap, and an MIT (Mott transition) occurs.

1.4 MAGNETIC PROPERTIES

TMOs display interesting magnetic properties in both naturally formed and artificially engineered materials, which is difficult to explain by conventional physics. In 1950, Anderson proposed a super-exchange model, where two next-nearest neighbour metal cations form an antiferromagnetic (AFM) coupling through a non-magnetic anion.[25] Goodenough and Kanamori propose magnetism in TMOs by super-exchange or double-exchange interaction.[26,27] TMOs exhibit exotic magnetic properties, for example, spin glasses, Griffiths phases, non-collinear spiral spin structures, and spontaneous magnetic reversal.[28-31] Slater, in 1951,[32] suggested that an electron is repelled much less strongly by other electrons of parallel spin than those of antiparallel spin. Subsequently, in the antiferromagnet phase, the doubling of the periodicity splits every band in two. A cubic crystal field again splits each of these sub-bands into two: a lower triply degenerate (t_{2g}) band and an upper doubly degenerate (e_g) band. For example, MnO, NiO, and CoO with partially filled *d*-electrons are still insulating in nature. These insulators also undergo small magneto-strictive distortions below their Neel temperatures. Moreover, above the Neel temperature, many of these materials show metallic behaviour. According to Slater's theory,[33] the exchange splitting is entirely dominated by the intra-ionic exchange and, therefore, should not vanish with the loss of long-range magnetic ordering. The most used procedure to understand the magnetic moment is by considering the localized magnetization and then applying super-exchange model, or the Heisenberg model, to determine the coupling constant.[34]

1.5 OPTICAL PROPERTIES

Optical properties show a material's response while exposing to electromagnetic radiation and in particular to visible light. TMOs, such as vanadium, scandium, and titanium oxides, display unique optical features due to the nonbonding nature of 3*d* orbitals with neighbouring cations. 3*d* orbitals of a particular TMO do not overlap with the oxides of other 3*d* metals and, therefore, maintain isolated energy states. For many years, optical measurements have been performed on TMOs. For example, in NiO, several optical absorption peaks are observed between 1 and 4 eV corresponding to the absorption by Ni^{2+} ions. Even in the presence of other oxides like MgO, the position of absorption peaks does not change.[35] TMO nanostructures and thin films exhibit colour absorption within the visible and UV region. Moreover, they exhibit photoluminescence due to their wide and controllable bandgap and, hence, are used as photochromic devices.[36,37] TMOs are widely used as thermochromic, as well as electrochromic materials, as they undergo reversible optical transformation as a function of temperature or electric field.[38,39] Thin films of WO_3 and V_2O_5 when exposed to hydrogen gas change their colour significantly and, therefore, are known as chemo-chromic material and are used for detecting hydrogen leakage.[40,41] The optical bandgap of TMO perovskite also ranges from 2 to 4 eV due to the large difference in electronegativity between the oxygen and metal cations. The optical bandgap in RMO_3 (R = rare earth element, M: transition metal), with M from Sc, Ti, V, Cr, Mn, Fe, Co, Ni, and Cu, shows a linear decrease from Sc to Cu.[42]

1.6 APPLICATIONS

The wide range of electrical, magnetic, and optical properties make the low-dimensional TMOs a promising material for nanoscience technological applications, where finite-size effects (quantum confinement) play an important role.[43] Multiple oxidation states, short diffusion path lengths, and large surface area make the nanostructured TMOs (MoO_2, WO_3, MnO_2, V_2O_5, NiO, and others) applicable as a cathode catalyst material in fuel cells.[44] Few TMOs also exhibit an electrochromic colouration effect and, therefore, are widely used as electrochromic devices.[39] Thin films of $LiMO_2$ (M: 3*d* transition metal) and $LiCoO_4$ find applications as energy storage devices, for example, rechargeable lithium-ion batteries,[45] Na-ion batteries,[46] and supercapacitors.[47] TMOs are also widely used as photocatalysts for the degradation of water, organic dyes etc., and photo-electrochemical cells[48,49] due to their mechanical and thermal stability, optical absorption in visible region, fast charge transfer capability, cost-effectiveness, and so on. Moreover, ultrathin MnO_2 nanosheets are used for gene delivery by absorbing double-stranded DNA, and nanostructured TiO_2 is used for efficient gene silencing.[50] We have discussed different applications of TMOs at low dimensions in detail in the respective chapters.

REFERENCES

1. C. N. R. Rao, *Annu. Rev. Phys. Chern.* 40 (1989) 291.
2. F. Pino, C. C. Mayorga-Martinez, A. Aerkoci, *Electrochem. Commun.* 71 (2016) 33.
3. A. Aboulouard, B. Gultekin, M. Can, M. Erol, A. Jouaiti, B. Elhadadi, C. Zafer, S. Demic, *J. Mater. Res. Technol.* 9 (2020) 1569.
4. S. Bhatia, N. Verma, *Mater. Res. Bull.* 95 (2017) 468.
5. M. S. Danish, L. L. Estrella, I. M. Alemaida, A. Lisin, N. Moiseev, M. Ahmadi, M. Nazari, M. Wali, H. Zaheb, T. Senjyu, *Metals* 11 (2021) 1.
6. C. C. Tseung, S. Jasem, *Electrochim. Acta* 22 (1977) 31.
7. S. Thambidurai, P. Gowthaman, M. Venkatachalam, S. Suresh, *J. Alloys Compd.* 830 (2020) 154642.
8. T. Jan, S. Azmat, Q. Mansoor, H. Waqas, M. Adil, S. Ilyas, I. Ahmad, M. Ismail, *Microb. Pathog.* 134 (2019) 103579.
9. M. Dutt, A. Ratan, M. Tomar, V. Gupta, V. Singh, *J. Phys. Chem. Solid* 145 (2020) 109536.
10. E. Salih, M. Mekawy, R. Y. Hassan, I. M. EI-Sherbiny, *J. Nanostructure Chem.* 6 (2016) 137.
11. Y. Zhang, X. Sun, L. Pan, H. Li, Z. Sun, C. Sun, B. K. Tay, *Solid State Ion.* 180 (2009) 1525.
12. N. Widiarti, J. Sae, S. Wahyuni, *Mater. Sci. Eng.* 172 (2017) 012036.
13. C. Ashok, K. V. Rao, C. S. Chakra, *J. Adv. Chem. Sci.* 2 (2016) 223.
14. Z. Durmus, B. Z. Kurt, A. Durmus, *Chem. Select* 4 (2019) 271.
15. A. Mehto, V. R. Mehto, J. Chauhan, I. Singh, R. Pandey, *J. Nanomed. Res.* 5 (2017) 1.
16. F. Ghoreishi, V. Ahmadi, M. Samadpour, *J. Nanostruct.* 3 (2013) 453.
17. V. E. Henrich, *The Surface Science of Metal Oxides*, Cambridge University Press, New York (2014) 14–61.
18. H. Kung, *Transition Metal Oxides: Surface Chemistry and Catalysis*, Elsevier, Amsterdam (2012).
19. H. Al-Abadleh, V. H. Grassian, *Surf. Sci. Rep.* 52 (2013) 63.
20. V. Henrich, *Surface and Near-surface Chemistry of Oxide Materials*, Elsevier, Amsterdam (1988) 23–30.
21. V. Sachan, D. J. Buttrey, J. M. Tranquada, J. E. Lorenzo, G. Shimne, *Phys. Rev. B* 51 (1995) 12742.
22. J. M. Tranquada, B. J. Sternlleb, J. D. Axe, Y. Nakamura, S. Uchlda, *Nature* 375 (1995) 561.
23. D. Adler, *Solid State Phys.* 21 (1968) 1.
24. N. F. Mott, *Phil. Mag.* 6 (1961) 287.
25. P. W. Anderson, *Phys. Rev.* 79 (1950) 350.
26. J. B. Goodenough, *Phys. Rev.* 100 (1955) 564.
27. J. Kanamori, *Prog. Theor. Phys.* 17 (1957) 197.
28. D. Khomskii, *Spin Electronics* (*eds.* M. Ziese, M. J. Thornton), Springer-Verlag, Berlin, Heidelberg 569 (2001) 89.
29. K.-Y. Meng, A. S. Ahmed, M. Baćani, A. O. Mandru, X. Zhao, N. Bagués, B. Esser, J. Flores, D. McComb, H. J. Hug, F. Y. Yang, *Nano Lett.* 19 (2019) 3169.
30. J. Noguesa, I. Schuller, *J. Magn. Magn Mater.* 192 (1999) 201.
31. M. Saghayezhian, S. Kouser, Z. Wang, H. Guo, R. Jin, Y. Zhu, J. Zhang, E.W. Plummer, *Proc. Nat. Acad. Sci. U.S.A.* 116 (2019) 10309.
32. J. C. Slater, *Phys. Rev.* 82 (1951) 538.
33. J. C. Slater, *J. Appl. Phys.* 39 (1968) 761.

34. C. Zener, *Phys. Rev.* 82 (1951) 403.
35. D. S. McClure, *J. Phys. Chem. Solids* 3 (1957) 311.
36. F. Babar, U. Mehmood, H. Asghar M. H. Mehdi, A. U. H. Khan, H. Khalid, N. U. Huda, Z. Fatima, *Renew. Sustain. Energy Rev.* 129 (2020) 109919.
37. G. H. Brown, *Photochromism*, John Wiley & Sons, New York (1971).
38. P. Monk, R. Mortimer, D. Rosseinsky, *Electrochromism: Fundamentals and Applications*, VCH, New York (2008).
39. P. Ashrit, *Electrochromic thin films and devices*, in *Transition Metal Oxide Thin Film Based Chromogenics and Devices* (*eds.* G. Korotcenkov), Elsevier, Amsterdam, Netherlands (2017) 73–151.
40. G. Korotcenkov, *Handbook of Gas Sensor Materials*, Springer, London 2 (2014) 181–184.
41. P. Liu, S. H. Lee, C. E. Tracy, J. A. Turner, J. R. Pitts, S. K. Deb, *Solid State Ionics* 165 (2003) 223.
42. T. Arima, Y. Tokura, J. B. Torrance, *Phys. Rev. B* 48 (1993) 17006.
43. M. Finazzi, L. Duo, F. Ciccacci, *Surf. Sci. Rep.* 64 (2009) 139. .
44. Z. Zhang, J. Liu, J. Gu, L. Su, L. Cheng, *Energy Environ. Sci.* 7 (2014) 2535.
45. R. Kanno, *Secondary batteries – Lithium rechargeable systems – Lithium-ion | positive electrode: Lithium nickel oxide*, in *Encyclopedia of Electrochemical Power Sources* (*eds.* J. Garche), Elsevier B. V., Amsterdam, Netherlands (2009) 297–306.
46. M. Mirzaeian, Q. Abbas, M. R. Hunt, A. Galeyeva, R. Raza, *Na-ion Batteries*, Elsevier B. V., Amsterdam, Netherlands (2021) 49.
47. S.-J. Park, Y.-R. Son, Y.-J. Heo, *Emerging Materials for Energy Conversion and Storage in Micro and Nano Technologies* (*eds.* K. W. Cheong. G. Impellizzeri, M. A. Fraha), Elsevier, Amsterdam, Netherlands (2018) 185–208.
48. S. R. Meher, *Nanostructured Materials for Visible Light Photocatalysis in Micro and Nano Technologies* (*eds.* A. K. Nayak, N. K. Sahu), Elsevier, Amsterdam, Netherlands (2022) 153–183.
49. S. Sankaranarayanan, M. Hariram, S. Vivekanandhan, C. Ngamcharussrivichai, *Green Functionalized Nanomaterials for Environmental Applications in Micro and Nano Technologies* (*eds.* U. Shanker, M. Rani, C. M. Hussain) Elsevier, Amsterdam, Netherlands (2022) 417–460.
50. M. Nurunnabi, J. McCarthy, *Biomedical Applications of Graphene and 2D Nanomaterials*, Elsevier, Amsterdam, Netherlands (2019).

2 Transition Metal Oxides at 2D

Structure, Properties, and Applications

2.1 INTRODUCTION

The discovery of exfoliated graphene has stimulated enormous research interest in other two-dimensional (2D) materials,[1] including transition metal oxide (TMO), transition metal chalcogenides, and transition metal carbides (nitrides), to realize novel functionalization and applications in semiconductor electronics.[2] Two-dimensional TMOs are an important group in the 2D family due to their structural diversity, cost-effectiveness, high stability, and unique optical, electrical, magnetic, and chemical properties.[3,4] Being an oxide, these 2D systems are not easily affected by air but easily lose oxygen or react with water and hydrogen.[5] The partially filled *d* orbitals of the transition metal ions and high electronegativity of oxygen atoms provide extraordinary electronic structures. Also, 2D oxides usually have a lower dielectric constant and a larger bandgap than their 3D counterparts, which make them applicable to be used in high-performance smart devices.[6,7] Perovskite TMO-based engineered heterostructures, discussed later in Chapter 5, have also attracted remarkable attention recently due to their emergent electronic and magnetic properties compared with those of their parent bulk compounds.[8]

In this chapter, we mostly focus on van der Waals (vdW)-bonded, and engineered 2D materials, their structural reconstruction, optical and electrical properties, and some potential applications.

2.1.1 2D MATERIALS: COMPOUND

Two-dimensional materials can generally be categorized as either 2D allotropes of single elements (e.g., graphene, borophene, silicene, etc.) or as compounds consisting of two or more covalently bonded elements (e.g., V_2O_5, TiO_2, WO_3, MoO_3, etc.). Two-dimensional compounds, especially TMOs, have the advantage of a highly active surface due to the unsaturated dangling bonds formed on the surface, which help them to exhibit many excellent physical and chemical properties.[9,10] However, most bulk TMOs have strong interlayer ionic bonds (except van der Walls TMOs), which makes it difficult to acquire their 2D monolayer by exfoliation. The surface of the active site is limited in the 3D bulk material, as only the outermost atoms can participate, whereas other atoms are covered. Hence, it is necessary to prepare an oxide monolayer

DOI: 10.1201/9781003623694-2

with compatible atomic thicknesses using bottom-to-top techniques that may lead to enhanced and adaptable optical, electrical, magnetic, and catalytic properties.

2.1.2 Monolayer and Its Perspectives

The reduction of layer thickness, especially in monolayer 2D materials, induces the quantum confinement effect that changes the electronic band structure, charge transport, and optical properties that are far different from the bulk materials.[11-14] The large surface-to-volume ratio and in-plane bound charge interaction in 2D TMOs open a wide range of tunings to structure,[15] hydrophilicity, hydrophobicity, and surface charge according to specific application needs.[16-18] The charge confinement in 2D TMOs provides unique electronic properties, whereas the monolayer thickness brings high mechanical flexibility and optical transparency.[19,20] Thus, monolayer 2D TMOs find their applications in flexible and transparent electronic and optoelectronic devices, which are small in size too. The easiest way to prepare monolayer TMO is by directly peeling a small portion of the 3D material that has a layered structure, either via mechanical exfoliation or by laser thinning.[21,22] However, it has been observed that the bottom-up growth approach (e.g., hydrothermal, solgel method, etc.) is more popular among scientists rather than the top-down method for synthesizing monolayer TMOs. The reason may be the strong chemical bonds between atoms in the 3D direction for some crystals. The monolayers can also be used to improve superlattice combinations, quantum wells, and complex heterostructures with unique electronic and magnetic properties and can be used as field-effect transistors, sensors, electronic switches, photochromic transducers, high-sensitivity detectors, and light modulators.[23-31]

2.2 VAN DER WAALS-BONDED 2D TRANSITION METAL OXIDES

There can be two types of 2D TMOs depending on the layer structure: (a) Van der Waals (vdW)-bonded, and (b) non-vdW bonded. The vdW-bonded 2D TMOs pose weak interactions between the planes, whereas non-vdW-bonded TMOs have atomic bonds between the planes with a limited number of layers. The schematic crystal structure diagrams of two well-known vdW-based TMOs are shown in Figure 2.1.[32] The recently developed 2D materials are mostly vdW-type layered compounds with strong in-plane covalent bonds but weak interplane interaction. The reason may lie in the synthesis procedure, as vdW-bonded 2D oxides can easily be separated from the 3D matrix by exfoliation techniques.[21,33] We will discuss a few well-known 2D vdW-bonded TMOs, that is, WO_3, MoO_3, and V_2O_5, their synthesis procedure, and emerging physical and chemical properties followed by their trending applications.

2.2.1 WO_3

WO_3 is a *n*-type wide bandgap semiconductor that finds applications in photo and electrochemical processes,[34] photocatalysts,[35] optoelectronics,[36] energy storage,[37] and gas sensing.[38] Especially 2D WO_3 shows excellent chemical and physical

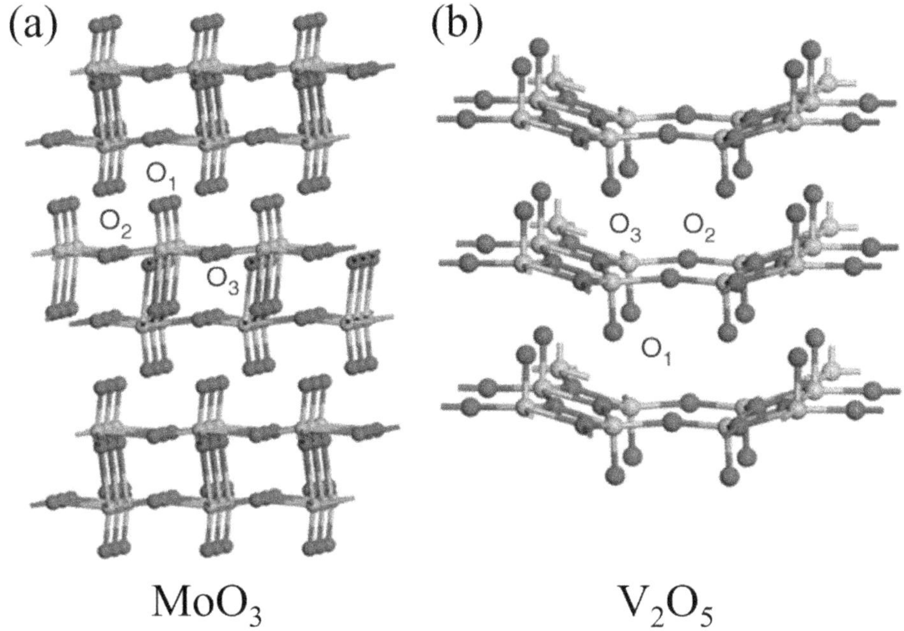

FIGURE 2.1 Crystal structure of orthorhombic (**a**) MoO_3 and orthorhombic (**b**) V_2O_5. (Reproduced with permission from Reference [32]. Copyright 2013, Wiley and 2015, WIREs Computational Molecular Science.)

properties with high specific surface area and high surface permeability, including electrochromism and photochromism, high electrical conductivity, and characteristic photoluminescence properties and, therefore, finds application in photodetectors,[39] light-emitting diodes (LED),[40] water splitting,[41] and so on.

2.2.1.1 Growth and Structural Characterization

2D WO_3 stabilizes in various crystallographic structures with corner- and edge-sharing WO_6 octahedra. The structures strongly depend on the octahedral distortion and little shift of the W atom.[42] However, the main difference among those closely structured phases relies on the fact that which W atom is displaced from the centre of WO_6 octahedra.[42] Therefore, the synthesis procedure plays a significant role in stabilizing a particular structure. Moreover, the transition among various structural phases can be made possible by altering the annealing temperature. A monoclinic-II (<−43°C) to monoclinic-I transition is observed by annealing at a temperature range between 17 and 330°C with an intermediate triclinic phase below 17°C.[43] A further annealing between 330 and 740°C is reported to produce orthorhombic β-WO_3.[44] Finally, a tetragonal α-WO_3 phase can be achieved with annealing at >740°C.[45] However, the stabilization of a new cubic phase is quite challenging and can only be possible via H intercalation.[46] There are several other ways to prepare WO_3, for example, via doping[47] or mechanical strain.[48] The well-known synthesis methods for

2D WO_3 are the solgel,[49] hydrothermal, and micro-emulsion methods.[50] Xiao *et al.*[51] synthesized nanosheets of WO_3 by hydrothermal route and used them for the degradation of pollutants. A microwave-assisted hydrothermal method was adopted by Perfecto *et al.*[52] to improve the sensing of isopropanol even at room temperature. Amano *et al.*[53] also used the hydrothermal method to grow 2D WO_3 on W plate. A thickness-controlled WO_3 platelet was obtained by Thind *et al.*[54] by varying the growth duration using hydrothermal method. There are also reports of obtaining 2D WO_3 thin films on Si wafers using thermal vapour deposition method.[55] Olejnícek *et al.*[56] used a magnetron sputtering system with W as a target in an Ar/O_2 environment to fabricate uniform WO_3 thin films. A 2D WO_3 atomic monolayer was deposited by Zhuiykov[57] by chemical vapour deposition (CVD) technique using $[(tBuN)_2(Me_2N)_2W]$ as a precursor.

2.2.1.2 Optical and Raman Spectroscopic Characterization

As described in the previous section, a little distortion can change the structural phase and, thereby, the optical and electrical properties of the 2D WO_3. Thus, it is important to know the local structure and change in bond vibrations with these distortions. Song *et al.*[58] have carried out a Raman spectroscopic analysis of the WO_3 nanosheets in its three well-known phases, namely monoclinic WO_3, triclinic WO_3, and hexagonal WO_3 as shown in Figure 2.2. They have observed characteristic Raman mode frequencies at 133 cm^{-1} (originated due to vibration of O–O bonds), 269, 319 cm^{-1} (bridging oxygen vibration of W–O–W bonds), and 709, 807 cm^{-1} (stretching vibration of the W–O–W bonds) for the mono- and triclinic-WO_3 nanosheets.[58]

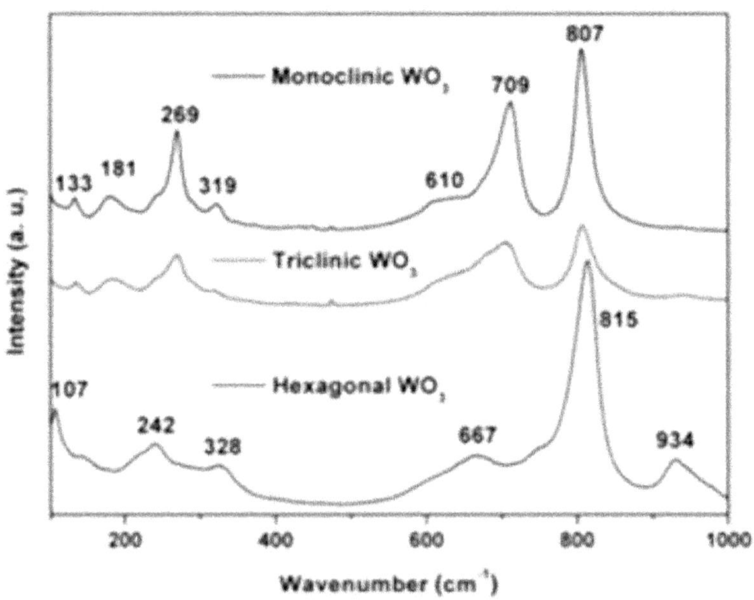

FIGURE 2.2 Raman spectra of monoclinic-, triclinic-, and hexagonal WO_3 nanosheets. (Reproduced with permission from Reference [58]. Copyright 2019, Springer Nature.)

A small Raman peak was also observed at 610 cm^{-1}, only in monoclinic WO_3, which corresponds to the stretching vibration of O–W–O bonds. However, the Raman spectra for hexagonal WO_3 are quite different than those for the other two phases of WO_3.[59] The highly intense Raman peak for hexagonal WO_3 is observed at 815 cm^{-1}, which is reported to originate due to the stretching vibration of O–W–O bonds. At lower energies, Raman mode frequencies of 242 and 328 cm^{-1} correspond to the bending vibration of W–O–W bonds, and the Raman mode frequency at 934 cm^{-1} corresponds to the bending vibration of W=O bonds of the hexagonal WO_3 phase.[59] Therefore, Raman spectroscopy can differentiate between the monoclinic, triclinic, and hexagonal phases of WO_3 by characteristic mode frequencies. Souza Filho *et al.*[60] performed a Raman spectroscopic investigation on WO_3 to check the phase coexistence and phase transition in the temperature range of 30–300 K.

2.2.1.3 Electrical Characterization

WO_3 is one of these few photocatalytic materials which operate in the visible range. It is non-toxic in nature and easy to fabricate. As the bandgap can be tuned between 2.4 and 3.2 eV,[61] an excellent absorption of the visible light makes it applicable as optoelectronic devices. The valence band of WO_3 is formed by filled O 2*p* orbitals with a potential energy of ~3 eV, whereas the conduction band is formed by empty W 5*d* orbitals.[61,62] However, as the structure and phase of WO_3 can be tuned by the degree of distortion of its structure, modulation of the conduction band and, thereby, the electronic bandgap can also be engineered accordingly. Amorphous WO_3 have a wide bandgap of ~3.2 eV. Whereas monoclinic WO_3 has a bandgap of ~2.6 eV.[62]

2.2.1.4 Applications

2.2.1.4.1 Electrochromism

WO_3 was the first discovered electrochromic material in 1815 by the Swedish chemist Jöns Jacob Berzeliusone and became the most extensively investigated chromogenic material thereafter.[63] Two-dimensional WO_3 nanosheets are the most suitable structure for device performance due to their large surface area, permeable channels, good surface attachment, and wide bandgap in the visible region, and therefore find applications as electrochromic displays and coloured memory devices, for example, smart windows, mobile display, reflectance mirrors, and eyewear.[64-69] The electrochromic effect can be tuned for a large frequency range by changing the wide bandgap through ion insertion/extraction.[70] The large electrochromic band also helps the material to be used as sensors, colour displays, and so on. However, there are conflicts on the origin of the colouration mechanism of WO_3-based electrochromic devices.[71] The mechanism is explained either by the optical transitions[71] or by collective electronic excitation in partially occupied bands[72] in the one-electron band structure. Moreover, the effects are often conferred as intervalence charge transfer in the case of 2D amorphous films.[73] Liang *et al.*[74] reported a coloration efficiency of 120.9 $cm^2 \cdot C^{-1}$ in ~1.4 nm thick WO_3 nanosheet. Hydrothermally grown porous phase pure and MoO_3-doped WO_3 films are reported to show colouration efficiency of 74.5 and 128.1 $cm^2{\cdot}C^{-1}$, respectively, with faster response time for the doped one.[75] A photovolta-chromic response from WO_3–perovskite film on a glass wafer was studied by Pugliese *et al.*[76] with a high light power efficiency of 12.2%.

2.2.1.4.2 Gas sensing

WO_3-based 2D thin films and nanosheets are also widely used as gas sensors mostly for H_2S, NH_3, H_2, CO, and other organic gases due to their large effective surface area, intrinsic oxygen vacancies, semiconducting properties, and easy fabrication.[77,78] Wang *et al.*[79] reported the sensing of NO_2 gas at a lower concentration of 2 ppb by WO_3 nanosheets grown via hydrothermal method with high selectivity and excellent response and recovery times. An increase in response with increasing temperature was observed up to 90°C, and the response decreased with a further increase in temperature. Yang *et al.*[80] investigated H_2 sensing by WO_3 nanoporous films at a range of 500 ppm limit for a temperature window between 200 and 300°C. A decreasing response with increasing temperature is observed for those WO_3 films grown via an electrochemical process using NaF/HF electrolytes. Leng *et al.*[81] studied the sensing response of 100 ppm of NH_3 at temperatures up to 500°C by WO_3 nanofibers synthesized via electrospinning method. A very high response of 5.5 was achieved at an operating temperature of 200°C, and the response and recovery times were ~1 and 5 s, respectively. Shen *et al.*[82] investigated the role of doping on the sensing properties of WO_3. Acetone is a selective biomarker for diabetes and, therefore, is being broadly studied for its sensing. Fe- and C-doped WO_3 show that at an optimum doping concentration of Fe (0.44 at %) and at an operating temperature of 300°C,[82] the response to acetone is superior with other gases.

2.2.2 MoO_3

One of the eminent vdW-bonded TMOs, MoO_3, stabilizes in three structures, namely, α-MoO_3, -, β (β', β'')-MoO_3, and hexagonal MoO_3 (Figure 2.3).[83,84] Among them, α-MoO_3 is thermodynamically stable and consists of layers along the *Y* axis. All the polymorphs of MoO_3 show unique physical and chemical properties, for example, refractive index, bandgap, and mechanical flexibility.[85] However, α-MoO_3 in the orthogonal phase shows enormous electrochemical and catalytic properties. The layer

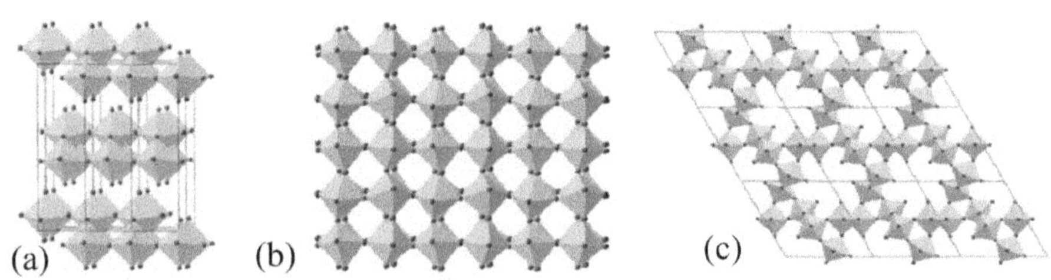

FIGURE 2.3 Schematic structures of MoO_3: (**a**) stable layered α-MoO_3, (**b**) cubic β-MoO_3, and (**c**) hexagonal h-MoO_3. (Reproduced with permission from Reference [84]. Copyright 2009, The Royal Society of Chemistry.)

of α-MoO_3 comprises corner-sharing MoO_6 octahedra connected by weak vdW forces in the *Y* direction (Figure 2.3a). Bulk α-MoO_3 has a large bandgap (~3 eV),[86] and exclusive optical and electrical properties, which result in its various technological applications. α-MoO_3 finds application in the fields of electrochromism and photochromism, photocatalysis, and solar cells.[87-90] Moreover, the multi-valency of Mo, rich thermal and chemical stability, and suitable redox chemistry make it applicable as gas sensor,[91] battery electrode,[92] and as a catalyst[93] for electrochemical reduction.

2.2.2.1 Growth and Structural Characterization

α-MoO_3 phase is constituted by corner-sharing MoO_6 octahedra along [001] and [100] directions, forming an orthorhombic structure with lattice parameters a = 3.963 Å, b = 13.86 Å, and c = 3.696 Å.[94] The anisotropic layers are stacked along the [010] direction with weak vdW forces. The extensive efficiency of 2D MoO_3 depends strongly on its morphology, structural revolution, and defect dynamics, indicating the importance of a controlled synthesis procedure. There are several reports on the controlled growth of MoO_3 thin films using precise techniques. Sian and Reddy[95] reported thermally deposited amorphous and crystalline thin films with rich optoelectronic properties. Mohamed *et al.*[96] prepared thin films using direct current (RF/DC) sputtering techniques. The electron beam evaporation (EBE) technique was used by Sivakumar *et al.*[97] with variation in the substrate temperature. Liu *et al.*[98] reported electrodeposited thin film on carbon electrodes. Chemical vapour deposited thin film was reported by Ivanova *et al.*[99] using $Mo(CO)_6$ precursor and Ar/O_2 gas mixture. Diskus *et al.*[100] used atomic layer deposition, and Gaigneaux *et al.*[101] used spin coating techniques to get the 2D MoO_3 films. Sun and Xiong[102] and Zheng *et al.*[103] reported solgel and hydrothermal methods, respectively, to synthesize MoO_3 nanosheets.

2.2.2.2 Optical and Raman Spectroscopic Characterization

vdW-bonded 2D α-MoO_3 undergoes a strong anisotropic phonon polariton effect in the middle- and far-infrared up to terahertz band[104] and, therefore, plays a crucial role in developing new photonics and optoelectronic devices at lower dimensions. Wen *et al.*[105] investigated the polarized Raman spectroscopic analysis on ~100 nm thick α-MoO_3 films by both theoretical and experimental analysis. They observed 14 Raman active modes among 24 groups theatrically predicted mode frequencies. The Raman peaks were observed at 85, 116, 128, 160, 191, 219, 245, 283, 338, 366, 441, 666, 820, and 996 cm^{-1}. Lee *et al.*[106] also studied the Raman spectroscopy of layered α-MoO_3 thin films. The layers of α-MoO_3 consist of corner-sharing chains of MoO_6 octahedra, in which three O atoms take part in shared edges of three octahedra, two O atoms are common in two octahedra, and one O remains unshared.[106] The authors observed three highly intense Raman peaks (Figure 2.4): the peak (i) at the highest frequency of 999 cm^{-1}, which is the characteristic mode for layer structure and is absent in other MoO_3 polymorphs, originated due to the vibration of $Mo^{6+}O$ bond (terminal unshared oxygen), (ii) at 821 cm^{-1} generated due to stretching vibration of Mo_2–O bond (common O in two octahedra), and (iii) at 671 cm^{-1} which corresponds to the stretching vibration of Mo_3–O bond (triply coordinated edge-shared O).

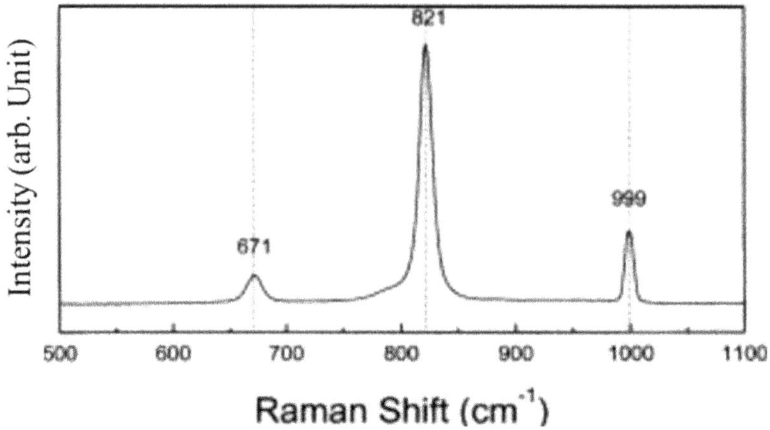

FIGURE 2.4 Raman spectrum of α-MoO_3 thin films. (Reproduced with permission from Reference [106]. Copyright 2002, Elsevier.)

2.2.2.3 Electrical Characterization

Layered α-MoO_3 is an *n*-type indirect semiconductor and has a large bandgap between 2.8 and 3.6 eV, which can be tuned via doping or by creating oxygen vacancy in the system.[107] Moreover, the layered nature helps it to manipulate the bandgap further by accumulating a huge number of positive ions in its tetrahedral and octahedral holes and extension channels, which leads to superior charge transfer.[107] The high ionic conductivity due to intrinsic oxygen vacancy and high stability makes it applicable to electronic devices.[108] The amount of oxygen vacancy 'x' in MoO_{3-x} plays a crucial role in the electrical conductivity, that is, for x between 0 and 0.125, it shows semiconducting nature, while for x between 0.125 and 1, it behaves as quasi-metallic, and for $x \sim 1$, it again becomes semiconducting with a reduced bandgap.[109] However, the presence of O vacancies changes the energy states of the parent system by increasing the concentration of carriers and forming a new donor level in the forbidden band.[110] The most popular and controlled way for creating O vacancy is by ion intercalation.[110] Wang *et al.*[111] used MoO_3 nanobelts as cathodes for Li-ion batteries (LIBs) with a capacity of 250 A•h•kg^{-1} in the potential window of 1.5-3.5 V by modifying its crystal structure and electrochemical properties. Du *et al.*[112] showed the Li^+ capacity of 700 mA•h•g^{-1} for 200 cycles for 2D α-MoO_3 nanosheets. Moreover, C-coated MoO_3 nanosheets are reported to show higher electrochemical performances owing to their synergetic effects.[113]

2.2.2.4 Applications

2.2.2.4.1 Electrochromism

The electrochromic properties of 2D MoO_3 TMOs have been studied with much attention due to their electrically induced discoloration properties. The layering shape with zigzag chains in the thermodynamically stable orthorhombic α-MoO_3 open channels is studied for ion diffusion and intercalation.[114] It also exhibits rich coloration efficiency, speedy response, and a small activation potential gradient. Moreover, MoO_3 shows more uniform and intense absorption of light as compared with WO_3, along with superior memory performance. MoO_3 nanobelts, synthesized by Lei *et al.*,[115] show rich photochromic and discoloration performance. Cai *et al.*[116] have prepared MoO_3 nanobelt arrays by morphology-controlled flame synthesis, exhibiting photochromic performances.

2.2.2.4.2 Gas Sensing

An appropriate gas-sensing material typically converts the analyte species information into electrical signals by means of a change in the material resistivity. Layered α-MoO_3 is a wide bandgap *n*-type semiconductor and has high structural stability and adaptable electrical conductivity. It, thus, has been considered as a sensor for several gases such as H_2S,[117] NO_2,[118] NO,[119] H_2,[120] CO,[121] CO_2,[122] C_2H_5OH,[123] NH_3,[124] trimethylamine,[125] acetone,[126] water/humidity sensor,[127] and glucose.[128] Bisht *et al.*[129] show sensing of NO_2 gas by α-MoO_3 thin film in ppm limit with 25% sensitivity at 100 C. α-MoO_3 nanosheets were reported to sense H_2S gas (10 ppm) with 38% sensitivity in the temperature range of 100–300 C.[117,130] CO_2 sensing was reported by Thomas *et al.*[122] with 38% sensitivity at 250 C. The NH_3 sensing was possible even at room temperature with high sensitivity at sub-ppm level as shown by Sakhuja *et al.*[131] However, surface modifications or decoration by various doping elements are reported to enhance the sensing properties by several orders.[132-135] Pt-decorated α-MoO_3 is reported to sense H_2 gas at 500 ppm with 80% sensitivity,[120] and 200 ppm HCHO with ~40% sensitivity at room temperature.[136] Whereas, Au-decorated α-MoO_3 were used for toluene and xylene detection at 100 ppm limit with ~20% sensitivity at 250 C.[137]

2.2.3 V_2O_5

Orthorhombic V_2O_5 has space group P_{mmn}, (D_{2h}^{13}), and it crystalizes by VO_5 pyramids sharing their vertices and corners.[138] The structure consists of weakly vdW-bonded anisotropic layers perpendicular to the *Z*-axis.[139] Among other vdW-bonded TMOs such as MnO_2 or MoO_3, V_2O_5 gained tremendous interest mostly due to its potential in applications such as gas sensing, electrochromic behaviour, optical and electrical switching, cathode materials for LIBs, and so on.[140-143] V_2O_5 is an important catalyst because of its three non-equivalent oxygen atoms in each unit cell (denoted as O_I, O_{II}, and O_{III} in Figure 2.5). O_I is the terminal (vanadyl) oxygen with two different bond lengths of 1.577 and 2.793 Å. The bridging oxygen (O_{II}) connects two adjacent V atoms with a bond length of 1.78 Å. The third O_{III} atom is ladder-like with V–O_{III} bond lengths of 1.88, 1.88, and 2.02 Å.[139] Previous research showed that O_I could easily be removed or added, leading to local changes in the vanadium oxidation state.[139] Unique electronic properties make V_2O_5 application in various fields.

2.2.3.1 Growth and Structural Characterization

V_2O_5 thin films are reported to be synthesized via several methods, including electron beam evaporation, pulsed laser deposition, chemical vapour deposition, solgel,

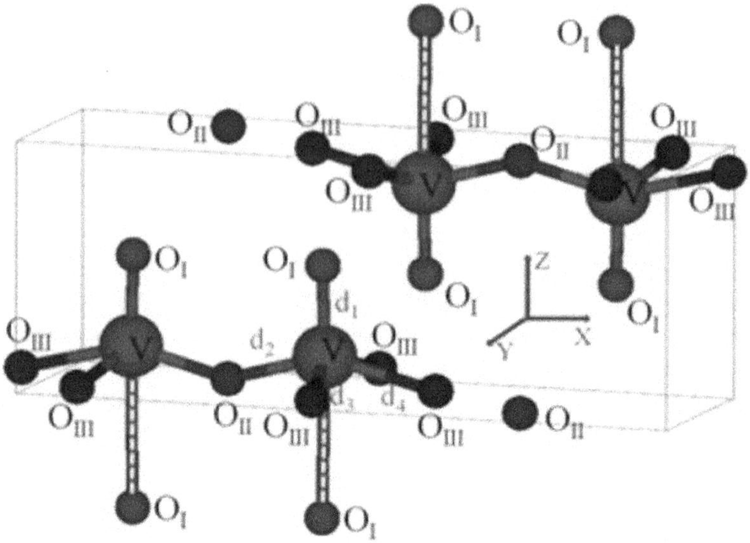

FIGURE 2.5 Schematic diagram of V_2O_5 unit cell. (Reproduced from Reference [139]. Creative Commons.)

spin coating, and sputtering. Moreover, 2D layers of vdW-bonded V_2O_5 can be chemically exfoliated for large-scale application of 2D layers. Radhakrishnan *et al.*[144] used the chemical exfoliation technique to synthesize the thinnest bilayer of 2D V_2O_5 nanosheets using formamide. Rui *et al.*[145] and Qi *et al.*[146] reported the synthesis of 2D V_2O_5 nanosheets using a liquid exfoliation technique. A gas-driven exfoliation of atomic-layered V_2O_5 was carried out by Wang *et al.*[147] Javed *et al.*[148] synthesized 2D V_2O_5 nanosheets via hydrothermal method.

2.2.3.2 Optical and Raman Spectroscopic Characterization

The optical properties of TMOs vary significantly with their dimension. Basu *et al.*[139] reported 11 Raman active modes at 102 (A_g), 144 (either B_{1g} or B_{3g}; B_{1g}/B_{3g}), 195 (A_g/B_{2g}), 283 (B_{1g}/B_{3g}), 301 (A_g), 403 (A_g), 483 (A_g), 523 (A_g), 701 (B_{1g}/B_{3g}), 850 (B_{1g}), and 993 (A_g) cm^{-1} among group theoretically predicted 21 ($7A_g+7B_{2g}+3B_{1g}+4B_{3g}$) Raman modes at Γ point (Figure 2.6). Orthorhombic V_2O_5 consists of distorted VO_5 pyramids which share their edges and corners.[138] The low-frequency modes at 102, 144, and 195 cm^{-1} are reported to originate due to the *X*, *Y*, and *Z* displacements (shear and rotations) of the whole ladder-like V-O_{III}-V chains.[149]

The intensity of the Raman mode at 144 cm^{-1} depends on the long-range ordering of V-O layers.[150] Oscillation of O_I atoms along the *X* and *Y* axes generates the mode frequencies at 403 and 283 cm^{-1}, respectively. The V-O_{II}-V bending vibration along the *Z* direction is responsible for the Raman peak at 483 cm^{-1}. Raman modes at 701 cm^{-1} (V-O_{III}-V antiphase stretching mode) and 523 cm^{-1} are reported to originate from the displacement of O_{III} atoms in the *Y* and *X* directions, respectively.[151] Raman peak at 301 cm^{-1} corresponds to the vibration of O_{II} atoms along the *Z* axis. The

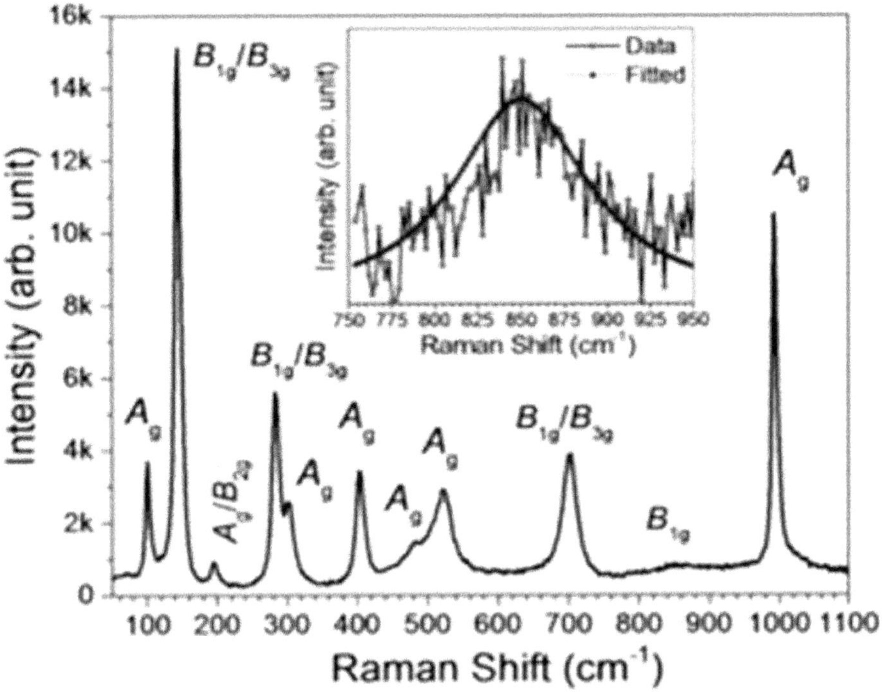

FIGURE 2.6 Raman spectrum of the as-grown V_2O_5 nanostructure at room temperature. The inset shows a fitted peak at 850 cm^{-1} corresponding to B_{1g} mode. (Reproduced from Reference [139]. Creative Commons.)

Raman mode frequency observed at the highest energy, 993 cm^{-1}, originated from the stretching vibration of the V-O_I bond along the *Z* direction.[152] The next peak at 850 cm^{-1} is generated due to the antiphase stretching mode of V-O_{II} bonds along the *X* direction.[139,149] A significantly increased (~7 times) optical absorption coefficient was reported in bilayer 2D vdW-bonded V_2O_5.[153]

2.2.3.3 Electrical Characterization

Unlike other vanadium oxides, such as VO_2 and V_2O_3, V_2O_5 is not believed to show a metal-to-insulator transition (MIT). However, researchers observed a reversible transition to a metallic phase of 2D V_2O_5 films around 230-253 C.[154] A surface metallicity of (001) facet around 40-100 C was also reported by Blum *et al.*[155] The metallic nature at higher temperatures helps V_2O_5 to find applications in thermochromic devices, gas sensors, optical and electrical switches, and so on.[153] The reason behind the metallic behaviour at this higher temperature is claimed to be due to the reduction of V_2O_5 to other lower-ordered oxides, which undergo a phase transition.[156] The metallic transition was also explained by the partial reduction of V_2O_5 and loss of oxygen from the structure. However, there was disagreement in pointing to the particular oxygen (O_I, O_{II}, or O_{III}) responsible for the vacancy creation.[157] The breakdown of $pd\pi$ bonds between vanadium and vanadyl O_I are finally reported to occupy the partially filled V *3d* bands by accumulating electrons toward neighbouring V atom, leading to the observed metallic behaviour in V_2O_5.[139]

2.2.3.4 Applications

2.2.3.4.1 Catalysis

Two-dimensional vdW-bonded V_2O_5 has bandgap in the range of ~2.3-3.7 eV, depending on the layer thickness and therefore has the potential to be used in the visible region without any modifications.[158] The multi-valency of V provides high oxidizing power and long-term chemical and thermal stability, which makes V_2O_5 an ideal candidate for photocatalyst.[153,159] In this section, we will mostly discuss the photo-degradation properties of 2D V_2O_5 for removing organic pollutants, for example, organic dyes, phenolic compounds, and pharmaceutical waste. Radhakrishnan *et al.*[153] reported ~ 99% degradation of organic dye at a limit of 16 ppm using the 2D V_2O_5 nanosheet under UV light exposure. Preeyanghaa *et al.*[160] reported the photo-degradation performance of dye pollutants using 2D g-C_3N_4/V_2O_5. Sekar *et al.*[161] reported a photocatalytic dye-degradation efficiency of 98.4% in 105 min for the graphitic carbon (GC)/ V_2O_5 structures in visible light. A degradation of 97.95% in 90 min in direct sunlight for Victoria blue dye was reported by Kamalam *et al.*[162] using graphene oxide V_2O_5/ GO nanostructures. In the same year (2021), Jenifer *et al.*[163] reported a degradation of 92% of methyl violet and 82% of methylene blue dyes using only V_2O_5 nanostructures. Oliveros *et al.*[164] used V_2O_5/BCN to degrade diclofenac (DCF) and achieved almost 100% degradation in less than 105 min. Photo-degradation of 88.35% for tetracycline (TC) by V_2O_5 and NH_2-MIL-101(Fe) was reported by Huang *et al.*[165] Liu *et al.*[166] explored ~100% photo-degradation of TC antibiotics using polymer-clothed TiO_2/ V_2O_5 nanobelts. Parashuram *et al.*[167] reported a 95% breakdown of bisphenol A (BPA) within 80 min using V_2O_5 decorated nitrogen-doped carbon.

2.2.3.4.2 Gas Sensing

Two-dimensional V_2O_5-based gas sensors demonstrate excellent responses to various target gases due to their high surface area and unique electrical and chemical properties. The sensing response strongly depends on the morphology and, therefore, on the synthesis procedure of the samples. The dependence of morphology in the V_2O_5 nanostructures prepared by magnetron sputtering was investigated by Mounasamy *et al.*[168] toward the sensing of methane gas. They have synthesized three different morphologies by varying the spurring power, namely, nanorods, nano-urchin, and nanoflower composed of nanosheets.[168] The resistance of the three different samples were ~15.4, 8.8, and 1.8 MΩ at room temperature (Figure 2.7a). A higher sensor response was observed for the nanoflower morphology (11.2%) than that of nanorods (8.9%) and nano-urchin (9.1%) at 100°C for 500 ppm CH_4 (Figure 2.7b-d). The reason behind the superior response was explained by the 2D structure of nanoflower composites by nanosheets, providing more active sites than that of the other two morphologies.

Mane and Moholkar[169] have studied the effect of film thickness on the sensing response. The nanocrystalline V_2O_5 thin films prepared by spray pyrolysis technique with thicknesses of 423, 559, 694, and 730 nm were used for NO_2 gas sensing. The V_2O_5 film with a thickness of 559 nm showed the highest response (41% to 100

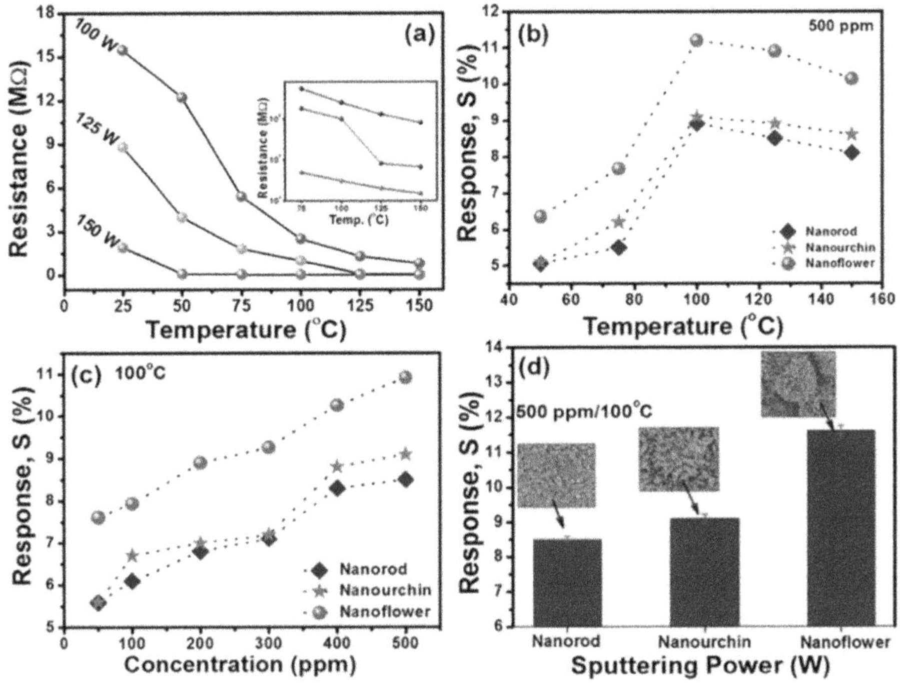

FIGURE 2.7 (**a**) Resistance versus temperature for V_2O_5 prepared under different sputtering powers, (**b**) response of different morphologies of V_2O_5 as a function of temperature, (**c**) response of different V_2O_5 nanostructures versus CH_4 concentrations at 100°C, (**d**) comparison between the responses to 500 ppm CH_4 of different morphologies of V_2O_5 at 100°C. (Reproduced with permission from Reference [168]. Copyright 2020, Elsevier.)

ppm NO_2 at 200°C) due to the higher surface area.[169] Nanoplates of V_2O_5 with less than 80 nm thickness show a high response to trimethylamine at 370°C.[170] Wang *et al.*[171] reported hydrogen gas sensing at room temperature by V_2O_5 nanostructures. At room temperature, sensing of acetone was also reported by Hakim *et al.*[172] using pristine V_2O_5. The V_2O_5 nanosheets exhibited a response (R_a/R_g) of ~2.8 for 100 ppm trimethylamine (TMA) with fast response/recovery times (5/28 s) at 240°C.[173] Meng *et al.*[173] also reported the response of 2.2 for 5 ppm TMA at 200°C with long-term stability and good selectivity using V_2O_5 thin films. Hydrothermally synthesized V_2O_5 nanosheets by Yang *et al.*[174] show three times higher response to 1-butylamine than that of commercial V_2O_5. The V_2O_5 nanosheets showed a response of 2.6-100 ppm 1-buthylamine with a fast response time of 6 s. Modafferi *et al.*[175] reported NH_3 sensing by V_2O_5 nanostructures synthesized by the electrospinning method.

2.2.3.4.3 Storage Cell

LIBs exhibit high energy density, long cycling life, insignificant memory effect, low self-discharge effect, and, most importantly, environmental friendliness and, therefore, are found to be the most promising electrochemical energy storage devices.[176]

However, a longer standby time, shorter charging time, and sustained service life are required for portable electronics.[177] Thus, an upgraded electrode material of LIBs is always crucial to achieve higher specific capacity and rich energy and power density along with longer cycling life.[178]

Vanadium oxides, especially with layered crystal structures, are considered one of the most capable electrode materials for next-generation energy storage technology due to their rich specific capacity, availability, and cost-effectiveness.[179] Almost all vanadium oxides, for example, V_2O_5, V_6O_{13}, VO_2(B), V_2O_3, and $V_3O_7{\cdot}H_2O$@C, along with vanadates (Li_3VO_4, $NaVO_2$, NaV_2O_5, etc.), are broadly used as electrode material for LIBs.[180] However, those materials with lower dimensions, especially at the nano-level, have shown superior properties because of the larger specific surface area and shorter ion diffusion path and are used for various energy storage applications. However, despite higher capacity, those nanostructured electrode materials suffer from serious self-aggregation and pulverization,[181] leading to poor cycling and rate performance. Moreover, these low-dimensional nanomaterials also exhibit less volumetric energy density due to their low tap density because of larger surface area. A promising way out to overcome these disadvantages is to prepare porous micro/nanostructures[182] consisting of nano-building blocks, which provide more surface area and excellent structural stability. For example, outstanding Coulombic efficiency, improved volumetric capacity, and fast cycling performance were observed in a plum-pudding-like Si/C micro/nano-composite prepared via quasi-industrial-scale production.[183] Similarly, porous reduced graphene oxide (rGO) and $LiFePO_4$/N composites exhibited excellent performance as cathodes for LIBs.[184] Wang *et al.*[185] have synthesized template-based MoS_2 micro-boxes embraced by ultrathin nanosheets, which show high specific capacity and first-rate cycling performance. However, free-standing electrodes with or without a supporting matrix are also an alternative choice to avoid the self-aggregation of nanomaterials.[186-188] The free-standing electrodes without any carbon black or polymer binder enhance the electrochemical performance.[189-191] Zhang *et al.*[189] prepared porous and highly dispersive Cu nanoarrays (CNAs) by *in situ* Cu foil method, which exhibited superior electrochemical performance (capacity over 640 $mA{\bullet}h{\bullet}g^{-1}$) as the free-standing anode of Na-ion batteries (NIBs). V_2O_5/PEDOT core/shell nanobelts array on 3D ultrathin graphite foam shows excellent electrochemical performance with a specific capacitance of 265 and 168 $mA{\bullet}h{\bullet}g^{-1}$ when cycled at 1.5 and 18 $A{\bullet}g^{-1}$, respectively.[192] It also maintained 98% retention up to 1000 cycles along with outstanding power and energy (238 $W{\bullet}h{\bullet}kg^{-1}$ at a power rate of 23 $kW{\bullet}kg^{-1}$) density.[193] Yang *et al.*[194] designed an interwoven nanocable architecture with V_2O_5 nanosheets encapsulated within multi-graphitic nanotubes by electrospinning method and used it as a novel and high-performance flexible cathode for LIBs.

2.2.3.4.4 Smart Window

Two-dimensional V_2O_5 also finds its application as smart windows due to its unique electrochromic properties, which range from optically transparent to optically dense state by tuning its optical bandgap. Due to this potential, V_2O_5 can be used as sunglasses, rear-view mirrors, and building windows. Alamri *et al.*[195] reported a thickness-dependent variation of coloration efficiency from 48 to 64 $cm^2{\bullet}C^{-1}$ when the thickness of V_2O_5 reduces from 313 to 78 nm. Dalavi *et al.*[196] investigated V_2O_5-CeO_2 composites as passive counter electrodes in smart windows. A rapid increase in coloration efficiency from 47 to 76 $cm^2 \cdot C^{-1}$ was observed when CeO_2-V_2O_5 was coupled as a counter electrode with WO_3. Wu *et al.*[197] doped a small percentage of Ti in V_2O_5 and observed that the electrochromic switching rate increases when utilized in smart windows.

2.3 ENGINEERED 2D TRANSITION METAL OXIDES

2.3.1 TiO_2

Among other TMOs, TiO_2 holds a distinct place because of its tunable structure, hierarchical architecture, substantial electrochemical properties, chemical stability, and ease of fabrication.[198,199] TiO_2 finds application in photocatalysis,[200] and as an electrode for LIBs.[201] A number of research groups are working on TiO_2 to tune its properties by refining the specific surface area, porosity, etc. In this section, we will discuss the different hierarchical 2D nanostructures of TiO_2, their properties, and a few technical applications, for example, photocatalysis, electrochromism, and solar cells.

2.3.1.1 Growth and Structural Characterization

TiO_2 stabilizes in four crystal forms: rutile (tetragonal), anatase (tetragonal), brookite (rhombohedral), and TiO_2-B (monoclinic). Figure 2.8 represents the different

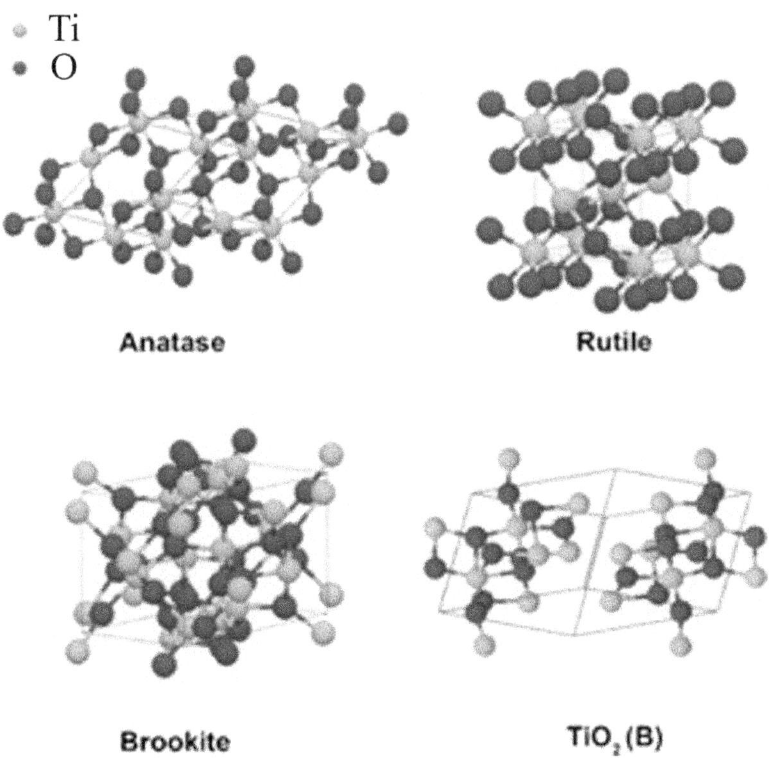

FIGURE 2.8 Different crystal structures of TiO_2. (Reproduced with permission from Reference [202]. Copyright 2021, Elsevier.)

structures of TiO_2. The monoclinic TiO_2-B can be transformed to tetragonal anatase at 800°C.[202] Brookite TiO_2 is the stable phase and shows superior photocatalytic activity than the metastable anatase TiO_2.[203] However, the stability also depends on the dimension; that is, in the bulk form the rutile phase is the most stable one, but the anatase and brookite phases are considered to be more stable in the nano-dimensions. In all four structural phases, a distorted TiO_6 octahedra exists. The corner-shared TiO_6 octahedra in the anatase phase changes to edge-shared octahedra in the rutile phase. In the brookite phase, both corners and edges of the TiO_6 octahedra are shared. However, the TiO_2-B phase has a more open structural framework, lower density, and a larger specific capacity.[204]

The 2D TiO_2 is generally synthesized by binding some facets strongly to the growing direction, thereby restricting the growth of crystals along one direction. A few popular ways to synthesize 2D TiO_2 are chemical vapour deposition,[205] microwave technique,[206] template-assisted synthesis,[207] solgel,[208] electrodeposition,[209] hydrothermal method,[210] etc. Han *et al.*[211] reported the synthesis of 2D nanosheets of TiO_2 by solvothermal method. Gopal *et al.*[212] prepared visible light active TiO_2 nanoflakes by heating the TiO_2 nanowires prepared with the hydrothermal method. Solgel-assisted pumice-TiO_2 nanoflakes were synthesized by Rad *et al.*,[213] which show excellent sonocatalytic activity. Sisman *et al.*[214] used TiO_2 nanoflakes for gas-sensing applications, which were grown by immersing titanium films in H_2O_2 on the alumina surface. The change in morphology and grain size of the TiO_2 nanoflakes due to calcination at high temperatures was analysed by Li *et al.*[215] Single-layered TiO_2 are generally prepared by delaminating protonic titanate.[216] The defect dynamics of the 2D TiO_2 nanosheets can be controlled using post-annealing in hydrogen and oxygen atmospheres.[217] Chen *et al.*[218] prepared a single-layered TiO_2 nanosheet by solvothermal route using glycerol and titanyl sulphate as precursors. Guo *et al.*[219] synthesized micrometre-long TiO_2 nanosheets on carbon fibres via a hydrothermal route. Exfoliation-mediated TiO_2 nanosheets were synthesized by Leng *et al.*[220] utilizing a surfactant.

2.3.1.2 Optical and Raman Spectroscopic Characterization

The optical properties of 2D TiO_2 depend on the phase present as well as the bandgap of the material. The three semiconducting phases of anatase, rutile, and brookite have a wide bandgap of ~3-3.2 eV, making it applicable for solar energy.[221] However, among the three natural phases of TiO_2, the anatase phase is the most important for application purposes. Group theory predicts six Raman active modes (A_{1g} + $2B_{1g}$ + $3E_g$) for the anatase phase (space group-$D^{19}/4h$), 144 cm^{-1} (E_g), 197 cm^{-1} (E_g), 399 cm^{-1} (B_{1g}), 513 cm^{-1} (B_{1g}), 519 cm^{-1} (A_{1g}), and 638.2 cm^{-1} (E_g).[222] The Raman mode frequencies are reported to have originated due to the symmetric, asymmetric, and bending vibration of Ti-O-Ti bond.[223] The reason behind the E_g modes is the stretching vibration of Ti-O. The Raman mode at 399 cm^{-1} corresponds to the O-Ti-O bending mode, whereas the mode frequency at 516 cm^{-1} indicates a Ti-O bond vibration due to the movement of O atoms. Zhang *et al.*[224] reported a thickness-dependent shift of E_g Raman mode in the 2D TiO_2 anatase phase, which offers a suitable and consistent method to analyse the thickness of the anatase TiO_2 nanosheets.

2.3.1.3 Electrical Characterization

TiO_2 semiconducting thin films have a wide energy gap of ~3.2 eV. However, an impurity-driven donor state below the conduction band is reported to be significantly above room temperature.[225] Above room temperature, the conductivity of the TiO_2 thin films is controlled by the thermally activated free carriers. However, below room temperature, the conductivity takes place through a variable range of hopping (VRH) between the localized states. The VRH may occur via either Mott physics[226] or Efros-Shklovskii (ES) law.[227] The electrical conductivity in these systems is mainly dominated by the hopping between impurity centres, rather than the free carriers, even at high temperatures. Yildiz *et al.*[228] showed that temperature dependences of electrical conductivity of the *n*-type TiO_2 thin films exhibit a crossover from Mott ($T^{1/4}$) to ES ($T^{1/2}$) dependence in the temperature range between 80 and 110 K.

2.3.1.4 Applications

2.3.1.4.1 Photocatalysis

Ultrathin 2D TiO_2 nanostructure with atomic-level resolution finds application in high-performance photocatalytic devices due to its large surface-to-volume ratio, characteristic electronic properties, and exciting chemical reactivity. However, the wide bandgap (3.2 eV for anatase and 3.0 eV for the rutile phase) of TiO_2 makes the conversion efficiency low. The main hurdle in TiO_2-based photocatalysts is the shifting of its optical response to the visible range.[229] The researchers are exploring methods to modify the absorption band by doping impurities in a controlled way. Doping of C,[230] N,[231] and S[232] is reported to lower its optical bandgap toward the visible light wavelengths. Another approach was also made by forming a 2D heterostructure of TiO_2 with a narrow bandgap semiconductor (M_xO_y/TiO_2 or M_xS_y/TiO_2). Previous research has demonstrated that TiO_2 doped with metallic ions helps widen the absorption range.[233] It also increases the redox kinetics and enhances the quantum efficiency by preventing the recombination of the electrons and holes (Figure 2.9). However, the photocatalytic activity also depends on the electronic properties and concentration of the dopant, as well as the length of the diffusion layer. Karakitsou and Verykios[234] investigated the role of valency and concentration of the doping metal (In^{3+}, Zn^{2+}, W^{6+}, Nb^{5+}, Li^+, etc.) on the photocatalytic performances. However, another aspect of dislocating defects formed by the dopant ions should also be taken care of to avoid the reduction of photocatalytic activity. Teh and Mohamed demonstrated that doping of rare earth metals in TiO_2 promotes catalytic properties by forming partially occupied 4*f* and empty 5*d* orbitals.[235] Various studies have extensively incorporated rare earth elements into the TiO_2 matrix to improve the photoactivity.[236] Dozzi *et al.*[237] and Kisch *et al.*[238] studied the noble metals (Pt and Au) doped TiO_2 photocatalyst for the degradation of formic acid and 4-chlorophenol under visible irradiation. The nanoparticles of noble metals increase the separation of photogenerated electron-hole pairs, thereby improving the photocatalytic properties. The coupling of TiO_2 thin films with other narrow bandgap semiconductors also gained tremendous attention. The photocatalytic activity is reported to be improved by several orders by coupling CdS, SnO_2, ZnO, WO_3, Bi_2S_3, and FeO_3 with TiO_2 for the irradiation of visible light.[239-245]

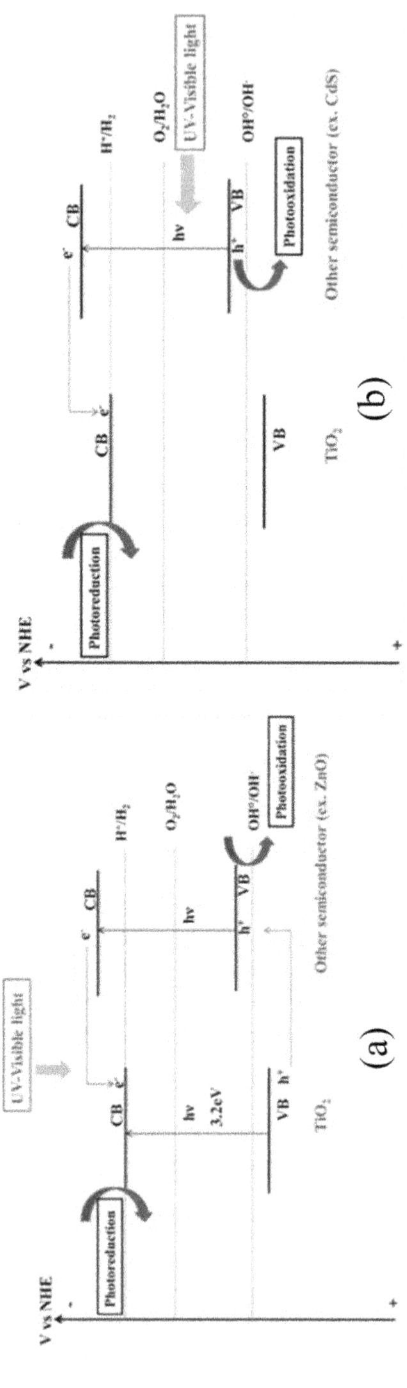

FIGURE 2.9 Vectorial transfer of (**a**) electrons and (**b**) holes in coupled semiconductor system. (Reproduced with permission from Reference [233]. Copyright 2013, American Chemical Society.)

2.3.1.4.2 *Electrochromism*

TiO_2 is used as an electrode material in semiconductor electrochemistry.[246] However, due to long coloration time and low coloration efficiency, TiO_2 shows poor electrochromic performance. Even though TiO_2 has gained strong attention as an electrochromic material because of its dual ability to work both as the cathodic and anodic electrochromic layers and multiple colour-changing properties.[247] The low electroconductivity and less ion diffusion coefficient in TiO_2 can be modified by improving its electronic properties by doping,[248] changing morphological dynamics,[249] and modulating crystal structure by reducing dimension.[250] Doping transition metal in 2D TiO_2 film is found to be the most fascinating method to enhance the electrochemical properties of TiO_2. The doping of transition metal ions improves the ion diffusion coefficient in TiO_2, as well as enhances the electroconductivity by variating the valence Ti state. A Ta-doped TiO_2 film shows enhanced electrochemical properties, as reported by Cao *et al.*[251] However, the optical modulation reduced by 1.3% after 2000 cycles at 550 nm. Niu *et al.*[252] doped Ce atom in TiO_2 film by solgel route and reported that an optimum doping of 2 mol% Ce improves the electrochemical properties by injecting a charge density up to 15.12 mC•cm^{-2}. However, doping non-transition elements is also reported to enhance the electrochromic properties by ion intercalation (H^+ and Li^+).[253] Cha *et al.*[254] and Ravishankar *et al.*[255] used Li^+-doped anatase TiO_2 to study the photo-electrochemical properties for applications such as solar cells and LIBs.

REFERENCES

1. K. S. Novoselov, A. K. Geim, S. V. Morozov, D. Jiang, Y. Zhang, S. V. Dubonos, I. V. Grigorieva, A. A. Firsov, *Science* 306 (2004) 5696.
2. A. K. Geim, I. V. Grigorieva, *Nature* 499 (2013) 419.
3. M. Osada, T. Sasaki, *J. Mater. Chem.* 19 (2009) 2503.
4. R. Ma, T. Sasaki, *Acc. Chem. Res.* 48 (2015) 136.
5. H. Zhang, *ACS Nano* 9 (2015) 9451.
6. P. Simon, Y. Gogotsi, *Nature Mater.* 7 (2008) 845.
7. W. Yang, L. Gan, H. Li, T. Zhai, *Inorg. Chem. Front.* 3 (2016) 433.
8. H. Jin, C. Guo, X. Liu, J. Liu, A. Vasileff, Y. Jiao, Y. Zheng, S.-Z. Qiao, *Chem. Rev.* 118 (2018) 6337.
9. C. S. Yang, D. S. Shang, N. Liu, G. Shi, X. Shen, R. C. Yu, Y. Q. Li, Y. Sun, *Adv. Mater.* 29 (2017a) 1.
10. T. Daeneke, P. Atkin, R. Orrell-Trigg, A. Zavabeti, T. Ahmed, S. Walia, M. Liu, Y. Tachibana, M. Javaid, A. D. Greentree, S. P. Russo, R. B. Kaner, K. Kalantar-Zadeh, *ACS Nano* 11 (2017) 10974.
11. P. Kumbhakar, C. Chowde Gowda, P. L. Mahapatra, M. Mukherjee, K. D. Malviya, M. Chaker, A. Chandra, B. Lahiri, P. M. Ajayan, D. Jariwala, A. Singh, C. S. Tiwary, *Mater. Today* 45 (2021) 142.
12. S. Z. Butler, S. M. Hollen, L. Cao, Y. Cui, J. A. Gupta, H. R. Gutiérrez, T. F. Heinz, S. S. Hong, J. Huang, A. F. Ismach, E. Johnston-Halperin, M. Kuno, V. V. Plashnitsa, R. D. Robinson, R. S. Ruoff, S. Salahuddin, J. Shan, L. Shi, M.G. Spencer, M. Terrones, W. Windl, J.E. Goldberger, *ACS Nano* 7 (2013) 2898.

13. M. Nasilowski, B. Mahler, E. Lhuillier, S. Ithurria, B. Dubertret, *Chem. Rev.* 116 (2016) 10934.
14. X. Kong, Q. Liu, C. Zhang, Z. Peng, Q. Chen, *Chem. Soc. Rev.* 46 (2017) 2127.
15. Y. Chen, Z. Fan, Z. Zhang, W. Niu, C. Li, N. Yang, B. Chen, H. Zhang, *Chem. Rev.* 118 (2018) 6409.
16. M. Yang, Z. Ye, M. A. Iqbal, H. Liang, Y.-J. Zeng, *Nanoscale* 14 (2022) 9576.
17. M. D. Stoller, S. Park, Y. Zhu, J. An, R. S. Ruoff, *Nano Lett.* 8 (2008) 3498.
18. T. Sasaki, M. Watanabe, H. Hashizume, H. Yamada, H. Nakazawa, *J. Am. Chem. Soc.* 118 (1996) 8329.
19. X. Cai, Y. Luo, B. Liu, H.-M. Cheng, *Chem. Soc. Rev.* 47 (2018) 6224.
20. G. Fiori, F. Bonaccorso, G. Iannaccone, T. Palacios, D. Neumaier, A. Seabaugh, S. K. Banerjee, L. Colombo, *Nat. Nanotechnol.* 9 (2014) 768.
21. C. Hermosa, B. R. Horrocks, J. I. Martínez, F. Liscio, J. Gómez-Herrero, F. Zamora, *Chem. Sci.* 6 (2015) 2553.
22. M. Yi, Z. Shen, *J. Mater. Chem. A* 3 (2015) 11700.
23. S. J. Chen, Y. C. Liu, C. L. Shao, R. Mu, Y. M. Lu, J. Y. Zhang, D. Z. Shen, X. W. Fan, *Adv. Mater.* 17 (2005) 586.
24. Y. Zhou, R. Ma, Y. Ebina, K. Takada, T. Sasaki, *Chem. Mater.* 18 (2006) 1235.
25. R. E. Schaak, T. E. Mallouk, *Chem. Mater.* 14 (2002) 1455.
26. R. E. Schaak, T. E. Mallouk, *Chem. Mater.* 12 (2000) 2513.
27. M. Osada, T. Sasaki, *J. Mater. Chem.* 19 (2009) 2503.
28. C. Felser, G. H. Fecher, B. Balke, *Angew. Chem. Int. Ed.* 46 (2007) 668.
29. M. Osada, Y. Ebina, K. Takada, T. Sasaki, *Adv. Mater.* 18 (2006) 295.
30. H. Sato, K. Ono, T. Sasaki, A. Yamagishi, *J. Phys. Chem. B* 107 (2003) 9824.
31. C. Tan, H. Zhang, *Nat. Commun.* 6 (2015) 7873.
32. M. Yang, Z. Ye, M. Iqbal, H. Liang, Y. Zeng, *Nanoscale* 14 (2022) 9576.
33. C. Wang, Y. Yi, H. Li, P. Wu, M. Li, W. Jiang, Z. Chen, H. Li, W. Zhu, S. Dai, *Nano Energy* 67 (2020) 104253.
34. Y. Ji, Y. Yang, S. K. Lee, G. Ruan, T. W. Kim, H. Fei, S. H. Lee, D. Y. Kim, J. Yoon, J. M. Tour, *ACS Nano* 10 (2016) 7598.
35. A. Fularz, S. Almohammed, J. H. Rice, *J. Phys. Chem. C* 124 (2020) 25351.
36. J. Sun, S. Zhang, T. Zhan, Z. Liu, J. Wang, X. Yi, J. Li, P. M. Sarro, G. Zhang, *J. Mater. Chem. C* 8 (2020) 5409.
37. T. Gaewdang, N. Wongcharoen, *Mater. Sci. Eng.* 383 (2018) 012006.
38. J. Dai, Y. Li, H. Ruan, Z. Ye, N. Chai, X. Wang, S. Qiu, W. Bai, M. Yang, *Nanomaterials* 11 (2021) 128.
39. Y. P. V. Karthik, B. Ajitha, Y. A. Kumar Reddy, V. R. Minnam Reddy, M. Reddeppad, M. D. Kim, *Appl. Surf. Sci.* 536 (2021) 147947.
40. L. Wang, S. Cheng, C. Wu, K. Pei, Y. Song, H. Li, Q. Wang, D. Sang, *Appl. Phys. Lett.* 110 (2017) 052106.
41. H. Lin, X. Long, Y. An, S. Yang, *J. Chem. Phys.* 152 (2020) 214704.
42. T. Vogt, P. M. Woodward, B. A. Hunter, *J. Solid State Chem.* 144 (1999) 209.
43. J. Z. Ou, M. S. Strano, R. B. Kaner, A. Mitchell, K. Kalantar-zadeh, H. Zheng, *Adv. Funct. Mater.* 21 (2011) 2175.
44. D. B. Migas, V. L. Shaposhnikov, V. N. Rodin, V. E. Borisenko, *J. Appl. Phys.* 108 (2010) 093713.
45. E. Boateng, S. S. Thind, S. Chen, A. Chen, *Electrochem. Sci. Adv.* 2 (2022) e2100146.
46. A. Hjelm, C. G. Granqvist, J. M. Wills, *Phys. Rev. B* 54 (1996) 2436.
47. M. M. Thwala, L. N. Dlamini, *Environ. Technol.* 41 (2019) 2277.

48. B. Cole, B. Marsen, E. Miller, Y. Yan, B. To, K. Jones, M. Al-Jassim, *J. Phys. Chem. C* 112 (2008) 5213.
49. B. Ji, W. F. Zhao, J. L. Duan, L. Z. Ma, L. H. Fu, Z. Yang, *J. Inorg. Mater* 35 (2020) 581.
50. R. Abazari, A. R. Mahjoub, L. A. Saghatforoush, S. Sanati, *Mater. Lett.* 133 (2014) 208.
51. Y. Xiao, Z. Q. He, R. S. Wang, X. Q. Tao, B. X. Li, *Colloids Surf. A* 580 (2019) 123752.
52. T. M. Perfecto, C. A. Zito, D. P. Volanti, *CrystEngComm* 19 (2017) 2733.
53. F. Amano, M. Tian, G. Wu, B. Ohtani, A. Chen, *ACS Appl. Mater. Inter.* 3 (2011) 4047.
54. S. S. Thind, M. Tian, A. Chen, *Electrochem. Commun.* 43 (2014) 13.
55. S. Li, Z. Yao, J. Yao, J. Zhou, R. Zhang, H. Shen, *Mater. Lett.* 195 (2017) 213.
56. J. Olejnícek, M. Brunclíková, Š. Kment, Z. Hubicka, H. Kmentová, P. Kšírová, M. Cada, M. Zlámal, J. Krysa, *Chem. Eng. J.* 318 (2017) 281.
57. S. Zhuiykov, L. Hyde, Z. Hai, M. K. Akbari, E. Kats, C. Detavernier, C. Xue, H. Xu, *Appl. Mater. Today* 6 (2017) 44.
58. W. Song, R. Zhang, X. Bai, Q. Jia, H. Ji, *J. Mater. Sci Mater. Electron.* 31 (2020) 610.
59. F. Zheng, M. Zhang, M. Guo, *Thin Solid Films* 534 (2013) 45.
60. A. G. Souza Filho, J. Mendes Filho, V. N. Freire, A. P. Ayala, J. M. Sasaki, P. T. C. Freire, F. E. A. Melo, J. F. Juliao, U. U. Gomes, *J. Raman Spectrosc.* 32 (2001) 695.
61. P. P. González-Borrero, F. Sato, A. N. Medina, M. L. Baesso, A. C. Bento, G. Baldissera, C. Persson, G. A. Niklasson, C. G. Granqvist, A. Ferreira da Silva, *Appl. Phys. Lett.* 96 (2010) 061909.
62. F. Wang, C. D. Valentin, G. Pacchioni, *ChemCatChem* 4 (2012) 476.
63. R. T. Wen, S. Malmgreen, C. G. Granqvist, G. A. Niklasson, *ACS Appl. Mater. Inter.* 9 (2017) 12872.
64. T. Varga, H. Haspel, A. Kormanyos, C. Janaky, A. Kukovecz, Z. Konya, *Electrochim. Acta* 256 (2017) 299.
65. X. Chen, S. S. Mao, *Chem. Rev.* 107 (2007) 2891.
66. H.-Y. Liu, C. C. Cody, J. A. Jacob-Dolan, R. H. Crabtree, G. W. Brudvig, *ACS Energy Lett.* 5 (2020) 3195.
67. D. S. Dalavi, R. S. Devan, R. A. Patil, R. S. Patil, Y. Ma, S. B. Sadale, I. Kim, J. H. Kim, P. S. Patil, *J. Mater. Chem. C* 1 (2013) 3722.
68. F. Zhao, X. Chen, Z. Yi, F. Qin, Y. Tang, W. Yao, Z. Zhou, Y. Yi, *Sol. Energy* 204 (2020) 635.
69. H. Najafi-Ashtiani, A. Bahari, S. A. Ghasemi, *J. Electroanal. Chem.* 774 (2016) 14.
70. S. Darmawi, S. Burkhardt, T. Leichweiss, D. A. Weber, S. Wenzel, J. Janek, M. T. Elm, P. J. Klar, *Phys. Chem. Chem. Phys.* 17 (2015) 15903.
71. N. Y. Bhosale, S. S. Mali, C. K. Hong, A. V. Kadam, *Electrochim. Acta* 246 (2017) 1112.
72. W. Han, Q. Shi, R. Hu, *Nanomaterials* 11 (2021) 692.
73. R. T. Wen, C. G. Granqvist, G. A. Niklasson, *Nat. Mater.* 14 (2015) 996.
74. L. Liang, J. Zhang, Y. Zhou, J. Xie, X. Zhang, M. Guan, B. Pan, Y. Xie, *Sci. Rep.* 3 (2013) 1936.
75. A. Hasani, Q. Van Le, T. P. Nguyen, K. S. Choi, W. Sohn, J. K. Kim, H. W. Jang, S. Y. Kim, *Sci. Rep.* 7 (2017) 13258.
76. M. Pugliese, F. Bisconti, A. Rizzo, S. Colella, C. T. Prontera, G. Gigli, V. Maiorano, P. Cossari, *ACS Appl. Energy Mater.* 2 (2020) 10453.
77. Y. S. Shim, H. G. Moon, D. H. Kim, L. H. Zhang, S. J. Yoon, Y. S. Yoon, C. Y. Kang, H. W. Jang, *RSC Adv.* 3 (2013) 10452.
78. R. Nowrouzi, F. Razi, F. Rahimi, A. I. Zad, *Sens. Lett.* 11 (2013) 2015.
79. C. Wang, R. Sun, X. Li, Y. Sun, F. Liu, G. Lu, *Sens. Actuator B Chem.* 204 (2014) 224.

80. T. Yang, Y. Zhang, Y. Cai, *J. Mater. Res.* 29 (2014) 166.
81. J. Leng, X. Xu, N. Lv, H. Fan, T. Zhang, *J. Colloid Inter. Sci.* 356 (2011) 54.
82. J. Y. Shen, M. D. Wang, Y. F. Wang, J. Y. Hu, Y. Zhu, Y. X. Zhang, Z. J. Li, H. C. Yao, *Sens. Actuator B Chem.* 256 (2018) 27.
83. C. Zhang, L. Zheng, Z. Zhang, R. Dai, Z. Wang, J. Zhang, Z. Ding, *Phys. Status Solid* 248 (2011) 1119.
84. N. A. Chernova, M. Roppolo, A. C. Dillon, M. S. Whittingham, *J. Mater. Chem.* 19 (2009) 2526.
85. L. Kihlborg, *Ark. Kemi* 21 (1963) 357.
86. R. L. Smith, G. S. Rohrer, *J. Solid State Chem.* 124 (1996) 104.
87. S. T. Wang, Y. G. Zhang, X. C. Ma, W. Z. Wang, X. B. Li, Z. D. Zhang, Y. T. Qian, *Solid State Commun.* 136 (2005) 283.
88. J. Wang, S. Dong, C. Yu, X. Han, J. Guo, J. Sun, *Catal. Commun.* 92 (2017) 100.
89. S. Ranjbar, G. Brammertz, B. Vermang, A. Hadipour, S. Cong, K. Suganuma, T. Schnabel, M. Meuris, A. F. da Cunha, J. Poortmans, *Phys. Status Solidi A* 214 (2016) 1.
90. X. Tan, L. Wang, C. Cheng, X. Yan, B. Shen, J. Zhang, *Chem. Commun.* 52 (2016) 2893.
91. C. Imawan, H. Steffes, F. Solzbacher, E. Obermeier, *Sens. Actuators B.* 78 (2001) 119.
92. K. Tang, S. A. Farooqi, X. Wang, C. Yan, *ChemSusChem* 12 (2019) 755.
93. H. Sinaim, D. J. Ham, J. S. Lee, A. Phuruangrat, S. Thongtem, T. Thongtem, *J. Alloy. Compd.* 512 (2012) 172.
94. Y. Zhu, Y. Yao, Z. Luo, C. Pan, J. Yang, Y. Fang, H. Deng, C. Liu, Q. Tan, F. Liu, Y. Guo, *Molecules* 25 (2019) 18.
95. T. S. Sian, G. B. Reddy, *Sol. Energy Mater. Sol. Cells* 82 (2004) 375.
96. S. H. Mohamed, O. Kappertz, J. M. Ngaruiya, T. P. L. Pedersen, R. Drese, M. Wuttig, *Thin Solid Films* 429 (2003) 135.
97. R. Sivakumar, R. Gopalakrishnan, M. Jayachandran, C. Sanjeeviraja, *Curr. Appl. Phys.* 7 (2007) 51.
98. S. Liu, Q. Zhang, E. Wang, S. Dong, *Electr. Commun.* 1 (1999) 365.
99. T. Ivanova, A. Szekeres, M. Gartner, D. Gogova, K. A. Gesheva, *Electrochim. Acta* 46 (2001) 2215.
100. M. Diskus, O. Nilsen, F. Helmer, *J. Mater. Chem.* 21 (2011) 705.
101. E. M. Gaigneaux, K. Fukui, Y. Iwasawa, *Thin Solid Films* 374 (2000) 49.
102. J. Sun, R. Xiong, *J. Sol. Gel Sci. Technol.* 27 (2003) 315.
103. L. Zheng, Y. Xu, D. Jin, Y. Xie, *J. Mater. Chem.* 20 (2010) 7135.
104. G. Hu, Q. Ou, G. Si, Y. Wu, J. Wu, Z. Dai, A. Krasnok, Y. Mazor, Q. Zhang, Q. Bao, C. W. Qiu, A. Alu, *Nature* 582 (2020) 209.
105. M. Wen, X. Chen, Z. Zheng, S. Deng, Z. Li, W. Wang, H. Chen, *J. Phys. Chem. C* 125 (2021) 765.
106. S.-H. Lee, M. J. Seong, C. E. Tracy, A. Mascarenhas, J. R. Pitts, S. K. Deb, *Solid State Ion.* 147 (2002) 129.
107. H. Gruber, E. Krautz, *Phys. Status Solidi A* 62 (1980) 615.
108. X. Zhao, W. Jia, X. Wu, Y. Lv, J. Qiu, J. Guo, X. Wang, D. Jia, J. Yan, D. Wu, *Carbon* 156 (2020) 445.
109. I. A. de Castro, R. S. Datta, J. Z. Ou, A. Castellanos-Gomez, S. Sriram, T. Daeneke, K. Kalantar-Zadeh, *Adv. Mater.* 29 (2017) 1701619.
110. Y. Cui, Y. Pu, Y. Hao, Q. Zhuang, *Russ. J. Electrochem.* 51 (2015) 119.
111. X. J. Wang, R. Nesper, C. Villevieille, P. Novák, *Adv. Energy Mater.* 3 (2013) 606.
112. Y. Du, J. He, G. Hou, F. Yuan, *Electrochim. Acta* 334 (2020) 135593.

113. X. Hou, M. Ruan, L. Zhou, J. Wu, B. Meng, W. Huang, K. Zhong, K. Yang, Z. Fang, K. Xie, *J. Energy Chem.* 78 (2023) 91.
114. S. Buller, J. Strunk, *J. Energy Chem.* 25 (2016) 171.
115. Z. Lei, X. Yang, J. Dong, X. Yi, *Chem. Mater.* 21 (2009) 5681.
116. L. Cai, P. M. Rao, X. Zheng, *Nano Lett.* 11 (2011) 872.
117. X. Li, H. Yang, X. Hu, Q. Wu, W. Xiong, Z. Qin, C. Xie, D. Zeng, *J. Hazard. Mater.* 429 (2022) 128270.
118. S. Bai, S. Chen, L. Chen, K. Zhang, R. Luo, D. Li, C. C. Liu, *Sens. Actuators B Chem.* 174 (2012) 51.
119. W. Li, Q. Ou, X. Wang, K. Xing, T. Tesfamichael, N. Motta, D.-C. Qi, *Appl. Surf. Sci.* 586 (2022) 152793.
120. S. Farzi-kahkesh, A. Fattah, M. B. Rahmani, *Microelectron. Eng.* 235 (2021) 111453.
121. J. Wang, Q. Zhou, Z. Wei, L. Xu, W. Zeng, *Ceram. Int.* 46 (2020) 29222.
122. T. Thomas, Y. Kumar, J. A. Ramos Ramón, V. Agarwal, S. S. Guzmán, R. Reshmi, S. Pushpan, S. L. Loredo, K. C. Sanal, *Sensors. Vacuum* 184 (2021) 109983.
123. H. Ji, W. Zeng, Y. Li, *Phys. E Low-Dimens. Syst. Nanostruct.* 114 (2019) 113646.
124. T. Thomas, N. Jayababu, J. Shruthi, A. Mathew, A. Cerdán-Pasarán, J. A. Hernández-Magallanes, K. C. R. Sanal, R. Reshmi, *Thin Solid Film.* 722 (2021) 138575.
125. S. Shen, X. Zhang, X. Cheng, Y. Xu, S. Gao, H. Zhao, X. Zhou, L. Huo, *ACS Appl. Nano Mater.* 2 (2019) 8016.
126. H. Fu, Z. Wu, X. Yang, P. He, X. An, S. Xiong, D. Han, *Appl. Surf. Sci.* 542 (2021) 148721.
127. H. Yu, C. Wang, F.-Y. Meng, J.-G. Liang, H. S. Kashan, K. K. Adhikari, L. Wang, E.-S. Kim, N.-Y. Kim, *Sens. Actuator B Chem.* 304 (2020) 127138.
128. A. M. Azharudeen, R. Karthiga, M. Rajarajan, A. Suganthi, *Microchem. J.* 157 (2020) 105006.
129. P. Bisht, A. Kumar, I. T. Jensen, M. Ahmad, B. D. Belle, B. R. Mehta, *Sens. Actuators B Chem.* 341 (2021) 129953.
130. F. Rahman, A. Zavabeti, M. A. Rahman, A. Arash, A. Mazumder, S. Walia, S. Sriram, M. Bhaskaran, S. Balendhran, *ACS Appl. Mater. Interfaces* 11 (2019) 40189.
131. N. Sakhuja, R. Jha, N. Bhat, *Mater. Sci. Eng. B* 271 (2021) 115249.
132. J. Zhang, P. Song, Z. Li, S. Zhang, Z. Yang, Q. Wang, *J. Alloys Compd.* 685 (2016) 1024.
133. J. Guo, H. Li, Q. Ma, Q. Zhang, Y. Wang, X. Wang, *Sens. Actuators B Chem.* 369 (2022) 132256.
134. F. Zheng, F. Dong, Z. Lv, H. Li, L. Zhou, Y. Chen, T. Huo, X. Luo, *Colloids Interface Sci. Commun.* 43 (2021) 100434.
135. W. Yang, B. Fang, X. Xiao, H. Meng, S. Liu, *Sens. Actuators B Chem.* 358 (2022) 131457.
136. X. Fu, P. Yang, X. Xiao, D. Zhou, R. Huang, X. Zhang, F. Cao, J. Xiong, Y. Hu, Y. Tu, Y. Zou, Z. Wang, H. Gu, *J. Alloys Compd.* 797 (2019) 666.
137. L.-L. Sui, X.-F. Zhang, X. Cheng, P. Wang, Y.-M. Xu, S. Gao, H. Zhao, L.-H. Huo, *ACS Appl. Mater. Interfaces* 9 (2017) 1661.
138. R. Enjalbert, J. Galy, *Acta Crystallogr. Sect. C Cryst. Struct. Commun.* 42 (1986) 1467.
139. R. Basu, A. K. Prasad, S. Dhara, A. Das, *J. Phys. Chem. C* 120 (2016) 26539.
140. I. Raible, M. Burghard, U. Schlecht, A. Yasuda, T. Vossmeyer, *Sens. Actuators B* 106 (2005) 730.
141. A. Dhayal Raj, T. Pazhanivel, P. S. Kumar, D. Mangalaraj, D. Nataraj, N. Ponpandian, *Curr. Appl. Phys.* 10 (2010) 531.
142. S. Nishio, M. Kakihana, *Chem. Mater.* 14 (2002) 3730.

143. L. Zuli, F. Guojia, W. Youqing, B. Yandong, *J. Phys. D Appl. Phys.* 33 (2000) 2327.
144. R. P. Radhakrishnan, P. Anees, S. K. Sinha, A. Dasgupta, G. Karuppiah, A. K. Prasad, S. Dhara, *J. Phys. Chem. Lett.* 12 (2021) 9821.
145. X. Rui, Z. Lu, H. Yu, D. Yang, H. H. Hng, T. M. Lim, Q. Yan, *Nanoscale* 5 (2013) 556.
146. Y. Qi, K. Qin, Y. Zou, L. Lin, Z. Jian, W. Chen, *Appl. Surf. Sci.* 514 (2020) 145950.
147. C. Wang, H. Li, X. Zhang, Y. Qiu, Q. Zhu, S. Xun, W. Yang, H. Li, Z. Chen, W. Zhu, *Energy Fuels* 34 (2020) 2612.
148. M. S. Javed, H. Lei, Z. Wang, B.-T. Liu, X. Cai, W. Mai, *Nano Energy* 70 (2020) 104573.
149. L. Abello, E. Husson, Y. Repelin, G. Lucazeau, *Spectrochim. Acta A Mol. Spectrosc.* 39 (1983) 641.
150. R. Baddour-Hadjean, J. P. Pereira-Ramos, C. Navone, M. Smirnov, *Chem. Mater.* 20 (2008) 1916.
151. P. Clauws, J. Broeckx, J. Vennik, *Phys. Stat. Sol. B* 131 (1985) 459.
152. W. Lambrecht, B. Djafari-Rouhani, J. Vennik, *Surf. Sci.* 126 (1983) 558.
153. R. P. Radhakrishnan, A. K. Prasad, S. Dhara, *Sci. Rep.* 14 (2024) 14462.
154. M. Kang, I. Kim, S. W. Kim, J.-W. Ryu, H. Y. Park, *Appl. Phys. Lett.* 98 (2011) 131907.
155. R.-P. Blum, H. Niehus, C. Hucho, R. Fortrie, M. V. Ganduglia-Pirovano, J. Sauer, S. Shaikhutdinov, H.-J. Freund, *Phys. Rev. Lett.* 9 (2007) 226103.
156. S. Xie, E. Iglesia, A. T. Bell, *J. Phys. Chem. B* 105 (2001) 5144.
157. R. Ramirez, B. Casal, L. Utrera, E. Ruiz-Hitzky, *J. Phys. Chem.* 94 (1990) 8960
158. S. K. Jayaraj, V. Sadishkumar, T. Arun, P. Thangadurai, *Mater. Sci. Semicond. Process.* 85 (2018) 122.
159. M. Shanmugam, A. Alsalme, A. Alghamdi, R. Jayavel, *ACS Appl. Mater. Interfaces* 7 (2015) 14905.
160. M. Preeyanghaa, V. Vinesh, B. Neppolian, *Chemosphere* 287 (2022) 132380.
161. S. Sekar, I. Rabani, C. Bathula, S. Kumar, S. Govindaraju, K. Yun, Y.-S. Seo, D. Y. Kim, S. Lee, *Environ. Res.* 205 (2024) 112201.
162. M. B. R. Kamalam, S. S. R. Inbanathan, K. Sethuraman, A. Umar, H. Algadi, A. A. Ibrahim, Q. I. Rahman, C. S. Garoufalis, S. Baskoutas, *Environ. Res.* 199 (2021) 111369.
163. A. Jenifer, M. L. S. Sastri, S. Sriram, *Optik (Stuttg)* 243 (2021) 167148.
164. A. N. Oliveros, J. A. I. Pimentel, M. D. G. de Luna, S. Garcia-Segura, R. R. M. Abarca, R. A. Doong, *Chem. Eng. J.* 403 (2021) 126213.
165. C. Huang, J. Wang, M. Li, X. Lei, Q. Wu, *Solid State Sci.* 117 (2021) 1.
166. J. Liu, S. Zhou, P. Gu, T. Zhang, D. Chen, N. Li, Q. Xu, J. Lu, *J. Colloid Interface Sci.* 578 (2020) 402.
167. L. Parashuram, M. K. Prashanth, P. Krishnaiah, C. B. P. Kumar, F. A. Alharti, K. Y. Kumar, B.-H. Jeon, M. S. Raghu, *Chemosphere* 287 (2022) 132348.
168. V. Mounasamy, G. K. Mani, D. Ponnusamy, K. Tsuchiya, P. R. Reshma, A. K. Prasad, S. Madanagurusamy, *Anal. Chim. Acta* 1106 (2020) 148.
169. A. A. Mane, A. V. Moholkar, *Appl. Surf. Sci.* 416 (2017) 511.
170. M. Wu, X. Zhang, S. Gao, X. Cheng, Z. Rong, Y. Xu, H. Zhao, L. Huo, *CrystEngComm* 15 (2013) 10123.
171. Y.-T. Wang, W.-T. Whang, C.-H. Chen, *ACS Appl. Mater. Interfaces* 7 (2015) 8480.
172. S. A. Hakim, Y. Liu, G. S. Zakharova, W. Chen, *RSC Adv.* 5 (2015) 23489.
173. D. Meng, J. Si, M. Wang, G. Wang, Y. Shen, X. San, F. Meng, *Chin. Chem. Lett.* 31 (2020) 2133.
174. T. Yang, H. Yu, B. Xiao, Z. Li, M. Zhang, *J. Alloys Compd.* 699 (2017) 921.

175. V. Modafferi, G. Panzera, A. Donato, P. L. Antonucci, C. Cannilla, N. Donato, D. Spadaro, G. Neri, *Sens. Actuators B Chem.* 163 (2012) 61.
176. J. M. Tarascon, M. Armand, *Nature* 414 (2001) 359.
177. M. N. Obrovac, V. L. Chevrier, *Chem. Rev.* 114 (2014) 11444.
178. X. Wang, Q. Weng, Y. Yang, Y. Bando, D. Golberg, *Chem. Soc. Rev.* 45 (2016) 4042.
179. A. Pan, H. B. Wu, L. Yu, X. W. Lou, *Angew. Chem. Int. Ed.* 52 (2013) 2226.
180. P. Liu, D. Zhou, K. Zhu, Q. Wu, Y. Wang, G. Tai, W. Zhang, Q. Gu, *Nanoscale* 8 (2016) 1975.
181. Y. Sun, H. Cheng, S. Gao, Z. Sun, Q. Liu, Q. Liu, F. Lei, T. Yao, J. He, S. Wei, Y. Xie, *Angew. Chem. Int. Ed.* 51 (2012) 8727.
182. P. Liu, Y. Xu, K. Zhu, K. Bian, J. Wang, X. Sun, Y. Gao, H. Luo, L. Lu, J. Liu, *J. Mater. Chem. A* 5 (2017) 8307.
183. X. Huang, H. Yu, J. Chen, Z. Lu, R. Yazami, H. H. Hng, *Adv. Mater.* 26 (2014) 1296.
184. G. Hou, B. Cheng, Y. Cao, M. Yao, B. Li, C. Zhang, Q. Weng, X. Wang, Y. Bando, D. Golberg, F. Yuan, *Nano Energy* 24 (2016) 111.
185. B. Wang, W. Al Abdulla, D. Wang, X. S. Zhao, *Energy Environ. Sci.* 8 (2015) 869.
186. L. Wen, F. Li, H. M. Cheng, *Adv. Mater.* 28 (2016) 4306.
187. S. Wan, J. Peng, L. Jiang, Q. Cheng, *Adv. Mater.* 28 (2016) 7862.
188. J. W. Lee, S. Y. Lim, H. M. Jeong, T. H. Hwang, J. K. Kang, J. W. Choi, *Energy Environ. Sci.* 5 (2012) 9889.
189. S. Yuan, X. L. Huang, D. L. Ma, H. G. Wang, F. Z. Meng, X. B. Zhang, *Adv. Mater.* 26 (2014) 2273.
190. Y. H. Zhu, S. Yuan, D. Bao, Y. B. Yin, H. X. Zhong, X. B. Zhang, J. M. Yan, Q. Jiang, *Adv. Mater.* 29 (2017) 1603719.
191. H. Gwon, H.-S. Kim, K. U. Lee, D.-H. Seo, Y. C. Park, Y.-S. Lee, B. T. Ahn, K. Kang, *Energy Environ. Sci.* 4 (2011) 1277.
192. D. Chao, X. Xia, J. Liu, Z. Fan, C. F. Ng, J. Lin, H. Zhang, Z. X. Shen, H. J. Fan, *Adv. Mater.* 26 (2014) 5794.
193. D. W. Su, S. X. Dou, G. X. Wang, *J. Mater. Chem. A* 2 (2014) 11185.
194. D. Kong, X. Li, Y. Zhang, X. Hai, B. Wang, X. Qiu, Q. Song, Q.-H. Yang, L. Zhi, *Energy Environ. Sci.* 9 (2016) 906.
195. S. N. Alamri, A. A. Joraid, *Mater. Sci. Forum* 663-665 (2010) 743.
196. D. Dalavi, A. K. Bhosale, R. Desai, P. S. Patil, *Mater. Today Proc.* 43 (2021) 2702.
197. I.-H. Wu, A. Chandrasekar, K. T. Arul, Y.-C. Huang, T. T. T. Nga, C.-L. Chen, J.-L. Chen, D.-H. Wei, K. Asokan, P.-H. Yeh, C.-H. Du, W.-C. Chou, C.-L. Dong, *Optical Mater. X* 22 (2024) 100301.
198. M. Fang, G. Dong, R. Wei, J. C. Ho, *Adv. Energy Mater.* 7 (2017) 1700559.
199. P. Thiyagarajan, H. Ahn, J. Lee, J. Yoon, J.-H. Jang, *Small* 9 (2013) 2341.
200. B. M. Reddy, A. Khan, *Catal. Rev. Sci. Eng.* 47 (2005) 257.
201. D. Ma, K. Li, J. H. Pan, *ACS Appl. Energy Mater.* 3 (2020) 4186.
202. S. Reghunath, D. Pinheiro, S. Devi, *Appl. Surf. Sci. Adv.* 3 (2021) 100063.
203. Q. Tay, X. Liu, Y. Tang, Z. Jiang, T. C. Sum, Z. Chen, *J. Phys. Chem. C* 117 (2013) 14973.
204. M. Altomare, M. Vittoria, G. Luca, A. Di, L. Palmisano, E. Selli, *Catal. Today*. 252 (2015) 184.
205. U. M. Patil, S. B. Kulkarni, P. R. Deshmukh, R. R. Salunkhe, C. D. Lokhande, *J. Alloys Compd.* 509 (2011) 6196.
206. F. Bondioli, A. M. Ferrari, C. Leonelli, C. Siligardi, G. C. Pellacani, *J. Am. Ceram. Soc.* 84 (2001) 2728.

207. B. M. S. Sander, M. J. Côtø, W. Gu, B. M. Kile, C. P. Tripp, *Adv. Mater.* 16 (2004) 2052.
208. K. Del Ángel-sanchez, O. Vázquez-cuchillo, A. Aguilar-elguezabal, A. Cruz-lópez, *Mater. Chem. Phys.* 139 (2013) 423.
209. J. Zhang, Y. Wang, C. Yu, X. Shu, L. Jiang, J. Cui, Z. Chen, T. Xie, Y. Wu, *New J. Chem.* 38 (2014) 4975.
210. Z. R. Tian, J. A. Voigt, J. Liu, B. Mckenzie, H. Xu, *J. Am. Chem. Soc.* 125 (2003) 12384.
211. X. Han, Q. Kuang, M. Jin, Z. Xie, L. Zheng, *J. Am. Chem. Soc.* 131 (2009) 3152.
212. N. O. Gopal, M. H. Basha, *Ceram. Int.* 44 (2018) 22129.
213. T. S. Rad, A. Khataee, B. Kayan, D. Kalderis, S. Akay, *J. Clean. Prod.* 202 (2018) 853.
214. O. Sisman, N. Poli, D. Zappa, E. Comini, *Proceedings 17th Int. Meeting Chem. Sensors* P2NG (2018) 880.
215. Y. Li, X. Han, J. Liang, X. Leng, K. Ye, C. Hou, K. Yu, *Chem. Res. Chin. Univ.* 31 (2015) 332.
216. L. Sheng, T. Liao, L. Kou, Z. Sun, *Mater. Today Energy.* 3 (2017) 32.
217. Q. Chen, H. Wang, C. Wang, R. Guan, R. Duan, Y. Fang, X. Hu, *Appl. Catal. B Environ.* 262 (2020) 118258.
218. Y. Chen, G. Tian, Z. Ren, C. Tian, K. Pan, W. Zhou, H. Fu, Eur. *J. Inorg. Chem.* (2011) 754.
219. W. Guo, F. Zhang, C. Lin, Z. L. Wang, *Adv. Mater.* 24 (2012) 4761.
220. M. Leng, Y. Chen, J. Xue, *Nanoscale* 6 (2014) 8531.
221. A.Fujishima, X. Zhang, D. A. Tryk, *Surf. Sci. Rep.* 63 (2008) 515.
222. T. Ohsaka, *J. Phys. Soc. Jpn.* 48 (1980) 1661.
223. A. Turkovic, M. Ivanda, S. Popovic, A. M. Tonejc, M. Gotic, P. Dubcek, S. Music, *J. Mol. Struct.* 410 (1997) 271.
224. W. Zhang, Y. He, M. Zhang, Z. Yin, Q. Chen, *J. Phys. D* 33 (2000) 912.
225. W. Ren, Z. Ai, F. Jia, L. Zhang, X. Fan, Z. Zou, *Appl. Catal. B* 69 (2007) 138.
226. F. Mott, *J. Non-Cryst. Solids* 1 (1968) 1.
227. A. L. Efros, B. I. Shklovskii, *Electric Properties of Doped Semiconductor*, Springer, Berlin (1984).
228. A. Yildiz, S. B. Lisesivdin, M. Kasap, D. Mardare, *J. Non-Cryst. Solids* 354 (2008) 4944.
229. Y. Zhang, X. Xiong, Y. Han, X. Zhang, F. Shen, S. Deng, H. Xiao, X. Yang, G. Yang, H. Peng, *Chemosphere* 88 (2012) 145.
230. S. U. M. Khan, M. Al-Shahry, W. B. Ingler Jr., *Science* 297 (2002) 2243.
231. R. Asahi, T. Morikawa, T. Ohwaki, K. Aoki, Y. Taga, *Science* 293 (2001) 269.
232. T. Ohno, T. Mitsui, M. Matsumura, *Chem. Lett.* 32 (2003) 364.
233. R. Daghrir, P. Drogui, D. Robert, *Ind. Eng. Chem. Res.* 52 (2013) 3581.
234. K. E. Karakitsou, X. E. Verykios, *J. Phys. Chem.* 97 (1993) 1184.
235. C. M. Teh, A. R. Mohamed, *J. Alloys Compd.* 509 (2011) 1648.
236. E. Grabowska, H. Remita, A. Zaleska, *Physicochem. Probl. Miner. Process.* 45 (2010) 29.
237. M. V. Dozzi, A. Saccomanni, E. Selli, *J. Hazard. Mater.* 211 (2012) 188.
238. H. Kisch, L. Zang, C. Lange, W. F. Maier, C. Antonius, D. Meissner, *Angew. Chem. Int. Ed.* 37 (1998) 3034.
239. J. C. Yu, L. Wu, J. Jin, P. Li, Q. Li, *Chem. Commun.* 9 (2003) 1552.
240. S. M. Gupta, M. Tripathi, *Chin. Sci. Bull.* 56 (2011) 1639.
241. B. B. Nayak, H. N. Acharya, G. B. Mitra, B. K. Mathur, *Thin Solid Films* 105 (1983) 17.

242. H. Tada, A. Kokubu, M. Iwasaki, S. Ito, *Langmuir* 20 (2004) 4665.
243. C.-F. Lin, C.-H. Wu, Z.-N. Onn, *J. Hazard. Mater.* 154 (2008) 1033.
244. M. Miyauchi, A. Nakajima, T. Watanabe, K. Hashimoto, *Chem. Mater.* 14 (2002) 4714.
245. R. Coehoorn, C. Haas, R. A. De Groot, *Phys. Rev. B* 35 (1987) 6203.
246. M. Lahav, M. E. van der Boom, *Adv. Mater.* 30 (2018) 1706641.
247. T. Ye, Y. Xiang, H. Ji, C. Hu, G. Wu, *RSC Adv.* 6 (2016) 30769.
248. I. Mjejri, M. Gaudon, A. Rougier, *Sol. Energy Mater. Sol. Cells* 198 (2019) 19.
249. K. R. Reyes-Gil, Z. D. Stephens, V. Stavila, D. B. Robinson, *ACS Appl. Mater. Inter.* 7 (2015) 2202.
250 J. Z. Chen, W. Y. Ko, Y. C. Yen, P. H. Chen, K. J. Lin, *ACS Nano* 6 (2012) 6633.
251. J. Cao, X. Song, X. Kang, Z. Dai, Z. Tan, *Adv. Powder Technol.* 29 (2018) 1925.
252. B. Niu, S. Wang, K. Wu, H. He, R. Zhang, *Materials* 11 (2018) 1910.
253. V. K. Thakur, G. Ding, J. Ma, P. S. Lee, X. Lu, *Adv. Mater.* 24 (2012) 4071.
254. G. Cha, S. Ozkan, I. Hwang, A. Mazare, P. Schmuki, *Surf. Sci.* 718 (2022) 122012.
255. T. N. Ravishankar, G. Nagaraju, J. Dupont, *Mater. Res. Bull.* 78 (2016) 103.

3 Transition Metal Oxides at 1D

3.1 QUASI-1D TMOS: CHARGE DENSITY WAVE

Density waves in solids represent two novel types of broken symmetry ground states of materials, namely, charge and spin density waves. These states evolve as the consequence of electron-phonon and electron-electron interactions in low-dimensional systems.[1] While the latter will be discussed in Chapter 7, the present section deals with charge density wave (CDW) in quasi-one-dimensional (1D) transition metal oxides (TMOs). Electron-phonon interactions leading to CDW are best referred to as a rigid ion approximation, where it is assumed that the ionic potential at any point depends only on the distance from the centre of the ion. Thus, the formalism of single-particle tunnelling in a superconductor can be implemented for describing the tunnelling in the CDW ground state. The second-order phase transition leading to the CDW ground state is examined by measurements of various thermodynamic quantities, namely, specific heat, thermal expansion, and elastic constants.

CDW was first envisaged in explaining metal-to-insulator transition (MIT) in heavily alkali metal-doped oxide, blue bronze ($K_{0.3}MoO_3$).[2] Optical measurements have shown for the first time that $K_{0.3}MoO_3$ is a 1D metal above the temperature of 180 K. The MIT at 180 K is understood as a Peierls transition,[3] leading to a pseudo gap of 0.15 eV in the density of states. The Peierls-type transition is associated with a tiny lattice distortion and associated CDW. Previously, Fogle and Perlstem[4] suggested a Mott-Hubbard transition as an explanation of the MIT at 180 K, though the 1D metallic nature of the blue bronze for $T > 180$ K was not realized at that time. The metallic state is unstable for 1D metal, and a second-order phase transition into an insulating state, known as Peierls-type transition,[3] is predicted. The foreseeable 3D coupling helps to induce MIT at finite temperatures.

A quasi-1D stripe domain of finite width was predicted in hole-doped superconducting cuprates of $La_{2-x}Sr_xCuO_4$.[5-7] Formation of domain walls of finite width, with the holes localized in bound states at the walls, is found to be energetically favourable. The prediction seems to be correlated to the reported anomalous suppression of superconductivity in $La_{2-x}Ba_xCuO_4$ and a few similar compounds with small but finite doping concentrations.[8-10] Incommensurable antiferromagnetic phase of CuO is separated by periodical domain walls infused with holes, and the process involves a dynamical modulation of spin and charge.[5-7,11] From the numerical

DOI: 10.1201/9781003623694-3

calculation, one can get an electronic structure where an additional mid-gap band pertaining to the quasi-1D stripe domain evolves along with the valence and conduction bands, separated by the antiferromagnetic gap. Each stripe domain unit cell donates one state to the band, and the quasi-1D stripe is optimally stable depending on whether the band is completely filled with the excess holes or not, for example, adding one more hole to the domain line in the uneven-uneven unit cell one can find that an additional spin bag is formed and removing a hole leads to a neutral unit cell in the line, both of which are unfavourable. A similar ordered quasi-1D stripe phase is reported in hole-doped La_2NiO_4.[12-14] La_2NiO_4 possess layered Ruddelsden–Popper-type structure having alternate La–Ni–O perovskite and La–O rocksalt layers along the *c*-axis.

A static correspondent of the dynamical stripe phase in $La_{1.6-x}Nd_{0.4}Sr_xCuO_4$, with $x = 0.12$, was observed using a neutron diffraction study of low-temperature tetragonal (LTT) structures (Figure 3.1).[15] A comparison of the low-temperature orthogonal (LTO) and LTT structures (Figure 3.2) is presented to make it clear why the latter phase is favourable for pinning a horizontal stripe phase. A diagonal pattern is formed by the atomic displacements in the LTO structure. On the other hand, a horizontal (or vertical) pattern is formed in the LTT case in the process. Thus, horizontal charge stripes are favourably pinned by the lattice potential in the LTT phase alone.

The observation was related to the anomalous suppression of superconductivity in the cuprate system[8-10] and suggested that superconductivity in the copper oxides is correlated to CDW. The CDW is also reported to be observed using variable-temperature scanning tunnelling microscopic (VT-STM) studies,[16] enabling simultaneous mapping of atomic-scale topography and differential conductance spectroscopy, proportional to the energy-dependent local density of states (LDOS) of the sample. The studies are performed on optimally doped $Bi_2Sr_2CaCu_2O_{8+\delta}$ (Bi-2212) with $T_c = 35$ K. A typical atomic-resolution STM topography (Figure 3.3a) shows a 785 Å region of the exposed BiO plane measured at 6 K with an inset showing the Pb (brighter) and Bi (dimmer) atoms. The CuO_2 plane lies approximately 5 Å below. A representative differential conductance spectrum from this area (Figure 3.3b) depicts a clear inner gap with peaks near 15 meV. The gap is possibly associated with the superconducting gap and a pseudo gap of 75 meV.[16] A differential conductance map of the region taken at a bias of 10 meV (Figure 3.3c) shows a checkerboard-like electronic lattice. The image is remarkably similar to those observed in other cuprates.[17-19] The checkerboard is observed to be beyond 50 meV at both positive and negative sample bias. The pattern, however, appears most strongly at low and positive bias. It appears in maps ranging from 10 to 300 mV (feedback set point voltages and feedback currents from 50 to 800 pA, and in topographic scans at 10 mV bias). The wavelength of this checkerboard is shown in Figure 3.3(d); the value is determined from the Fourier transform (FT) of the image. The checkerboard appears as four peaks (one being *circled*). Its wave vector corresponds to a wavelength $d \sim 6.2a_0 \pm 0.2a_0$, which is significantly larger than that of any such structure reported earlier.

Figure 3.3e shows a line cut along the atomic lattice $(\pi,0)$ direction of the FT-LDOS maps taken at different bias voltages. The consistent position of the checkerboard wave vector is observed at different energies, marked by the left *dashed line*,

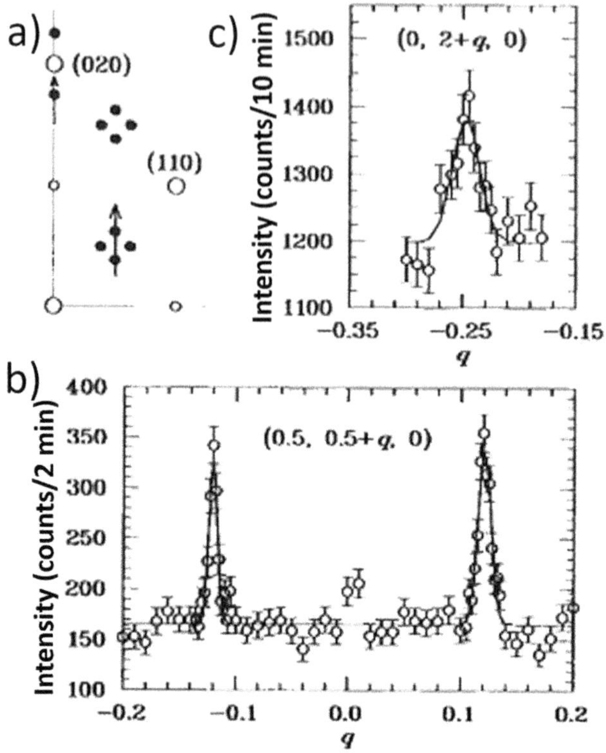

FIGURE 3.1 Scans of superlattice peaks consistent with the proposed spin and charge stripes in $La_{1.48}Nd_{0.4}Sr_{0.12}CuO_4$ at 11 K. (**a**) Diagram of the (hk0) zone of reciprocal space. *Open circles* indicate locations of Bragg peaks for the LTT structure and *solid circles* denote spin- and charge-ordering superlattice peaks. *Arrows* indicate the regions scanned. (**b**) Scan along (½, ½+q, 0) through the (½, ½±ε, 0) peaks measured with a neutron energy of 13.9 meV. The small peak width indicates that the in-plane correlation length is greater than 150 Å. (**c**) Scan along (0, 2+q, 0) through the (0, 2-2ε, 0) peak using 14.7 meV neutrons. The *lines* in (**b**) and (**c**) are the result of least squares fits to Gaussian peak shapes plus a flat background. (Reproduced from Reference [15]. Creative Commons.)

indicating a non-dispersive, static ordering persists in the checkerboard. VT-STM measurements exhibited a non-dispersive ‘checkerboard’-like electronic modulation that persisted in a broad regime of the cuprate phase diagram and showed a strong doping dependence (Figure 3.4). The continuous increase of checkerboard periodicity with hole density supported a strong correlation of checkerboard appearance with the CDW formation in the antinodal region of the cuprate Fermi surface.[20] A coherent picture of static electronic orderings in the cuprates emerges, emphasizing the nature of the pseudogap phase.

There is another class of CDW associated with the onset of ferromagnetic order. In this context, the ferromagnetism at T_{c3} = 175 K in $BaIrO_3$ seems to be associated

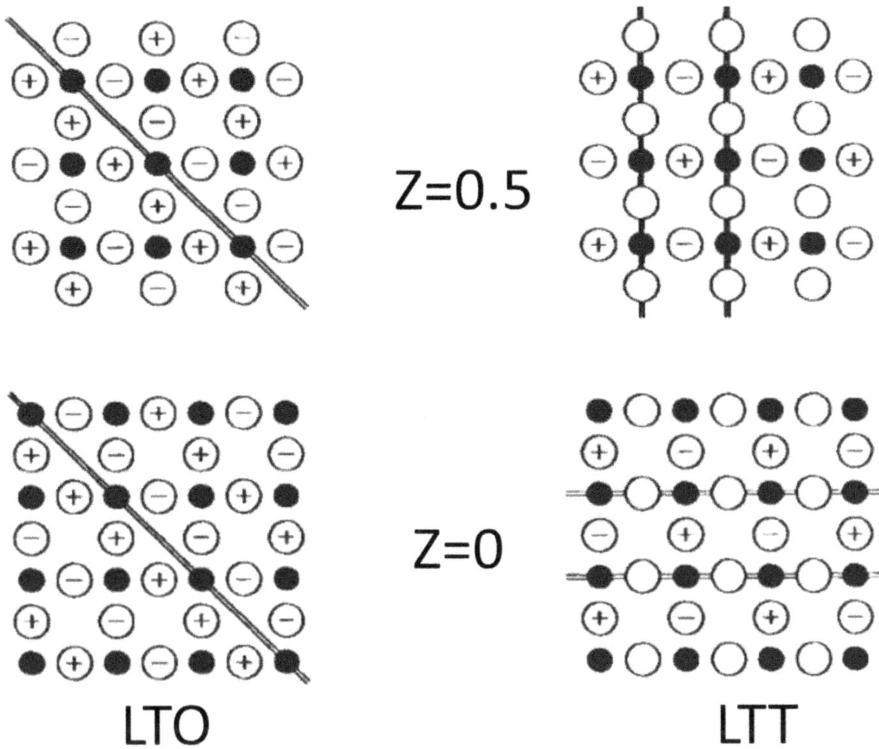

FIGURE 3.2 Displacement patterns within the two CuO_2 layers of a unit cell for the LTO and LTT structures. *Open* (*solid*) circles represent O (Cu) atoms. The oxygen atoms are displaced out of the plane (+ or -) by local rotations of square planar CuO_4 units about tilt axes shown by *doubled lines*. In the LTT structure, the tilt axes rotate by 90 from $z = 0$ to $z = 0.5$, where z is the height along the c-axis in lattice units. (Reprodued from Reference [15]. Creative Commons.)

with and perhaps driven by a collective electronic excitation or formation of a partial gap in the Fermi surface.[21] The observation reveals again the strong coupling between spin and charges in the 4*d*- and 5*d*-based heavy TMOs.[22-24] $BaIrO_3$ possesses an anisotropic structure with the quasi-1D *c*-axis having significantly low resistivity and shows a large anisotropy of temperature-dependent electrical resistivity. This kind of low-dimensional structure is necessary for the formation of an insulating CDW ground state, which is a collective electron mode normally incommensurate with the underlying lattice for partially filled bands.[1] Highly non-linear DC conductivity with two features in the chain direction shows a sudden change to metallic behaviour below $T_{c2} = 80$ K and then a Mott-like transition at $T_{c1} = 26$ K. Optical gap formation at <9 $k_B T_{c3}$, appearance of low-frequency phonon modes, and development of X-ray satellite structure support formation of CDW. Simultaneous structural and magnetic transitions are also reported in some 4*f* systems, where the magnetic ordering is driven by aspherical Coulomb scattering.[25] The abrupt features at T_{c1} and T_{c2} may be a consequence of the charge-ordered state originating from small polaron formation

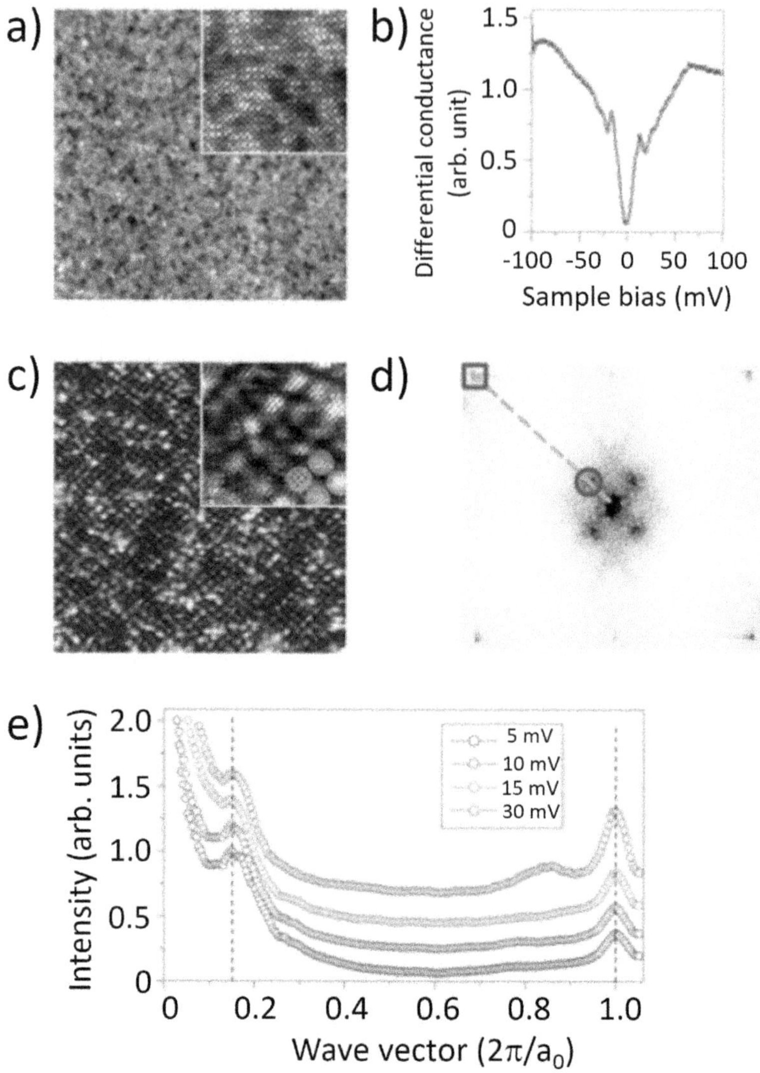

FIGURE 3.3 Anatomy of the checkerboard in optimally doped Bi-2201. (**a**) STM topography of a 785 Å region of optimally doped ($T_c = 35$ K) Bi-2201 measured at $T = 6$ K. The magnified inset (110 Å) (and *encircled* on nine atoms) show the clear atomic lattice in this high-resolution data. (**b**) Spatially averaged differential conductance spectra measured in the area shown in an exhibit as two distinct gaps: a superconducting gap Δ_{SC} ~15 meV and pseudogap Δ_{PG} ~75 meV. (**c**) Conductance maps, here taken with bias voltage of 10 meV on the same region as (**a**), show a checkerboard structure in the LDOS with a wavelength much larger than the atomic lattice. The inset is magnified as in (**a**), with the same nine atoms highlighted in *encircled* dark spots (red). Four checkerboard maxima are also highlighted (*circular* patches (yellow) adjacent to *encircled* dark spots) for clarity. (**d**) Fourier transform of the map shown in (**c**). The checkerboard wave vectors (*circled*) appear as four spots along the same direction as the atomic

lattice (*outlined with a square*). The *dashed line* shows the locations of the line cuts in (**e**). (**e**) Line cuts extracted from Fourier-transform LDOS maps with different bias voltages. The *left vertical line* marks the position of the checkerboard wave vector, $2\pi/6.2a_0$, for all energies, and the *right vertical line* indicates the atomic lattice wave vector $2\pi/a_0$. All data in this letter was acquired with feedback setpoint parameters IS = 400 pA and VS = −100 mV or VS = −200 mV. (Reproduced from Reference [20]. Creative Commons.)

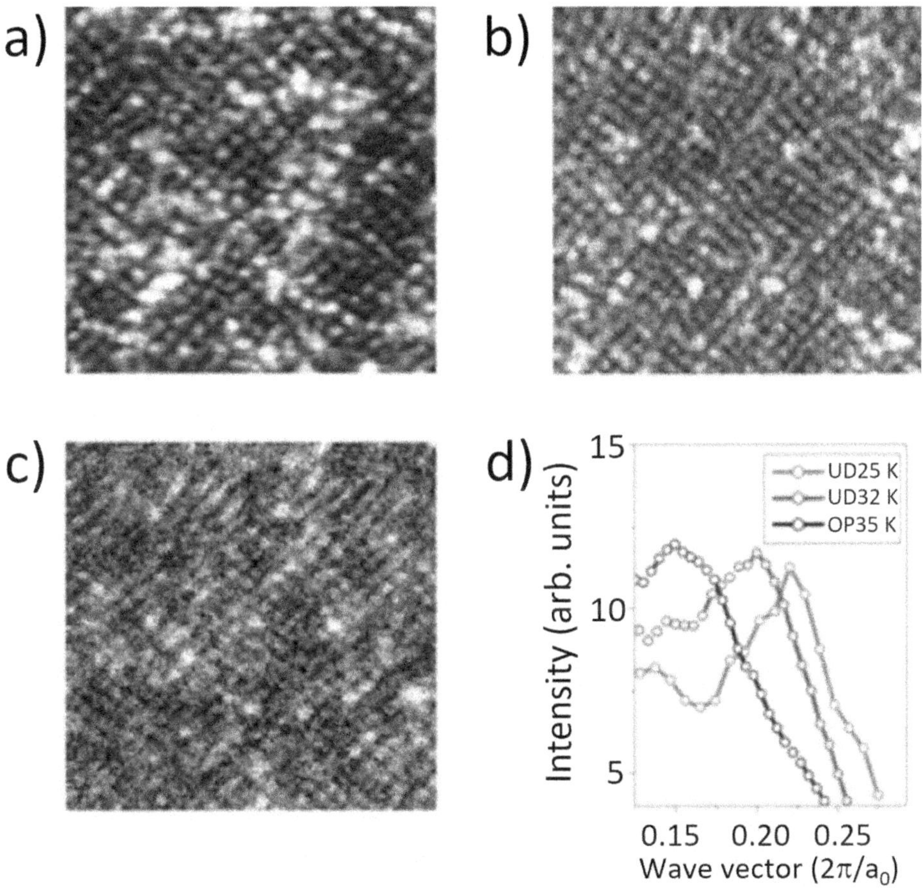

FIGURE 3.4 Doping dependence of the checkerboard. (**a-c**) The 400 Å conductance maps of optimally doped, $T_c = 35$ K (**a**); under-doped, $T_c = 32$ K (**b**); and under-doped, $T_c = 25$ K (**c**) Bi-2201. All maps were taken with 10 mV sample bias at $T = 6$ K. The checkerboard structures shown in (**b**), (**c**) have denser packing than in (**a**), indicating a shorter wavelength in under-doped samples. (**d**) Line cuts along the atomic lattice direction of the Fourier-transform LDOS maps of the three samples. The cuts peak at the checkerboard wave vectors, corresponding to wavelengths of $6.2a_0$, $5.1a_0$, and $4.5a_0$, respectively. (Reproduced from Reference [20]. Creative Commons.)

and/or CDW formation in this low-dimensional system and may be addressed as a 'small polaron CDW'.[26] The T_{c1} = 26 K anomaly adds $BaIrO_3$ to the special group of materials, demonstrating a spontaneous temperature-driven MIT. Similar to the organic charge transfer salts, the formation of CDW is anisotropic, with the 26 K transition being absent for the current perpendicular to the *c*-axis. The correlation between the onset of ordered magnetism and the formation of CDW in 4*d* and 5*d* TMOs merits further study.

3.2 ENERGY APPLICATIONS

3.2.1 Electrochemical Energy Storage

As discussed in Section 2.2.3.4.3, there are tremendous possibilities for vanadium-based oxides as electrode materials for next-generation electrochemical energy storage (EES) applications. Nanostructures of these materials have shown specific advantages for various EES applications owing to their ultra-large specific surface area and ultra-short diffusion path of carriers.[27] Nevertheless, nanostructured electrode materials face some problems,[28-30] namely, self-aggregation and pulverization during cycling because of significantly high large surface area and energy, leading to poor cycling and rate performance.[30,31] As an example, hydrothermally synthesized centimetre-long V_2O_5 nanowires demonstrated a high capacity of ~350 $mA{\times}h{\times}g^{-1}$ with capacitance retention of 50% after 20 cycles at a current rate of 50 $mA{\times}g^{-1}$, as cathode materials for lithium-ion batteries (LIBs).[32] The high capacity was attributed to the larger specific surface area of a 1D nanostructure. Cucumber-like layer-by-layer V_2O_5/poly(3,4-ethylenedioxythiophene) (PEDOT)@MnO_2 nanowires showing an initial capacity of ~180 $mA{\times}h{\times}g^{-1}$ at a current rate of 50 $mA{\times}g^{-1}$, with capacitance retention of 93% after 40 cycles was reported to improve the cycling performance of V_2O_5 nanowire (Figure 3.5).[33] However, it may be noted that the capacity of coated V_2O_5 nanowires decreased because of the presence of low-capacity PEDOT and MnO_2. Moreover, the capacity stability of composite V_2O_5 nanowires also needs to be further improved when cycled at a large current density.[33,34] Fabrication of free-standing electrodes with or without supporting matrix, meeting the demand in smart wearable electric, is a method to avoid the self-aggregation of nanowires.[35-38]

Recently, ultrathin V_2O_5 nanowires incorporated into uniformly graphene paper, making V_2O_5/graphene paper electrode,[39] showed a robust and extraordinary cycling performance with a capacity of ~192 $mA{\times}h{\times}g^{-1}$ being maintained when cycled at a current rate of 1 $A{\times}g^{-1}$ for 3000 cycles. A capacity of ~94 $mA{\times}h{\times}g^{-1}$ was also reported when cycled at an even higher current rate of 10 $A{\times}g^{-1}$ for very long (10^5) cycles. C. Niu *et al.*[40] studied hollow microspheres self-assembled by monoclinic VO_2 (B) nanowires for EES applications. As-synthesized VO_2 (B) (one of the polymorphs of VO_2) hollow microspheres exhibited high specific capacity of 200 and ~160 $mA{\times}h{\times}g^{-1}$ at current densities of 100 and 1000 $mA{\times}g^{-1}$, respectively, and improved cycling stability of 90% and 73% after 100 cycles at current densities of 10^2 and 10^3 $mA{\times}g^{-1}$, respectively. Balogun *et al.*[41] reported fascinating carbon cloth-supported carbon quantum dot surface-engineered VO_2 (C-VOCQD) interwoven nanowires as free-standing cathodes of LIBs. Pristine C-VOCQD demonstrated exceptional

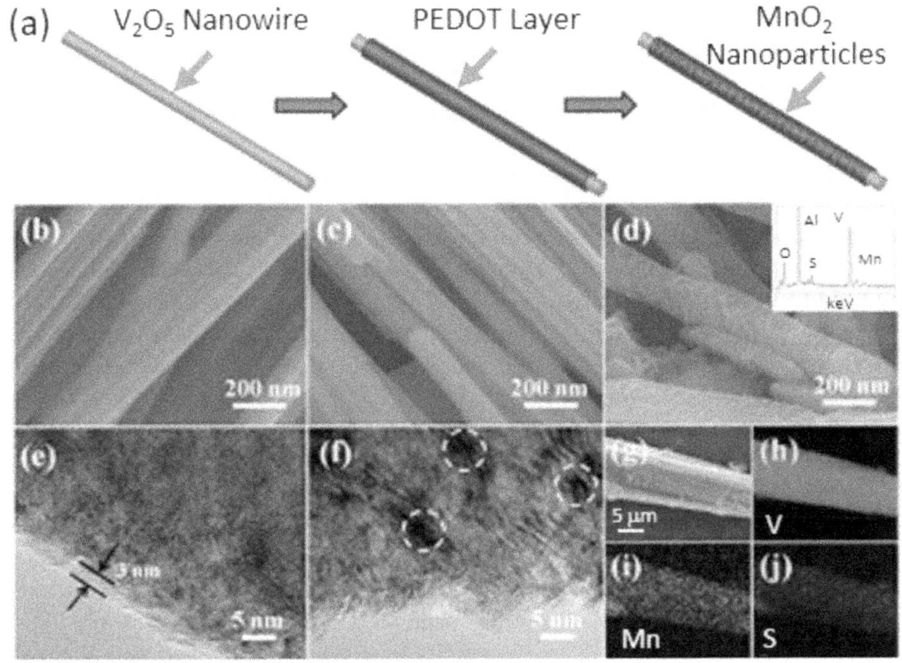

FIGURE 3.5 Characterizations of cucumber-like V_2O_5/PEDOT and MnO_2 nanowires: (**a**) schematic diagram of the synthesis of V_2O_5/PEDOT and MnO_2 nanowires; FESEM images of (**b**) V_2O_5 nanowires, (**c**) V_2O_5/PEDOT nanowires, and (**d**) V_2O_5/PEDOT and MnO_2 nanowires (the inset is the EDS spectra); TEM images of (**e**) V_2O_5/PEDOT nanowires and (**f**) V_2O_5/PEDOT and MnO_2 nanowires; FESEM image (**g**) and EDS mapping of V (**h**), Mn (**i**), and S (**j**) from V_2O_5/PEDOT and MnO_2 nanowires. (Reproduced with permission from [33]. Copyright 2013, American Chemical Society.)

electrochemical performance of ~400 $mA \times h \times g^{-1}$ at a current rate of 0.3 $A \times g^{-1}$. A capacity of ~170 $mA \times h \times g^{-1}$ was delivered at a very high current rate of 60 $A \times g^{-1}$ with more than 100% capacity retention after 500 cycles.

The 1D bimetallic spinel $Ni_xCo_{3-x}O_4$ electrode materials are recently reported for high-performance hybrid supercapacitor applications.[42] The unique nanowire-like architecture is advantageous for facilitating profuse bare active sites on large surface areas, leading to easy carrier transport conduits. As compared with their single counterparts, $Ni_xCo_{3-x}O_4$ electrode with an optimum ratio of Ni/Co = 1:1 is reported to show the maximum specific capacitance of 1033 $F \times g^{-1}$ at a current rate of 1 $A \times g^{-1}$ with the capacity retention of ~75% at a current rate of 30 $A \times g^{-1}$.[2] Moreover, the as-assembled hybrid supercapacitors integrating optimally grown $Ni_xCo_{3-x}O_4$ (Ni/Co = 1:1) and AC electrodes demonstrated large energy and power densities of ~42 $W \times h \times kg^{-1}$ and ~ 45 $kW \times kg^{-1}$ with excellent capacity retention value of 96.12%. Hydrothermally grown nanowire arrays of MnO_2@$NiCo_2O_4$(NCO) exhibiting a specific capacitance value of 452.1 $C \times g^{-1}$ at a current rate of 10 $A \times g^{-1}$ is reported for EES studies.[43] A reasonably good capacitance retention value of ~82% after 10^4 cycles was achieved with an energy density of ~32 W×h×kg^{-1} at a power rate of 2.2 kW×kg^{-1}.

A new technique for the design and synthesis of TMO hierarchical nano-arrays is reported for EES applications.[44] Compared with specific capacitance of ~640 F×g^{-1} at a current rate of 2 A×g^{-1} of the supercapacitor electrodes based on Co_3O_4 nanowire arrays, Co_3O_4/NiO nanowire arrays on 3D macroporous nickel foam, because of the unique porous core/shell nanowire array architecture, exhibited a high specific capacitance of ~850 F×g^{-1} and an impressive cycling stability after 6000 cycles. Up to 85% of capacitance is retained for the Co_3O_4/NiO core/shell nanowire arrays when the current rate is varied from 2 to 40 A×g^{-1}. The value is significantly higher than that for the Co_3O_4 nanowire arrays (76.7%). The discharge areal capacitance value of 1.35 F×cm^2 was reported for the core/shell nanowire arrays and found to be about twice larger than that for the Co_3O_4 nanowire arrays. EES properties of LIBs depend on the micro-nano features of active materials as well as electrode conformations. In this regard, the EES performance of mesoporous Ni-Mn-Co-O oxide nanowire arrays directly grown on Cu substrates as an integrated electrode for LIBs was reported.[45] The electrochemical measurements show that the nanowire-array-integrated electrodes without binder deliver enhanced cycling stability and high specific capacity compared with those for the electrode prepared by a conventional coating process. Moreover, the nanowire-array-integrated electrodes made on porous copper demonstrate excellent (~99%) cycle stability and rate capability of 0.2 A×g^{-1} compared with that on a Cu-foil base.

The 3D core@shell heterostructures consisting of highly conductive 1D nanowires as the core and electroactive TMO as shells and branches are used for EES devices. Using a sacrificial reactive 3D carbon template, Liu *et al.*[46] reported electrodes of Co_3O_4@MnO_2 nanowire arrays for a high value of capacitance with a reasonable cycle performance and significant rate capability, comparable to the best reported data for hybrid insulating metal oxide systems. In another study, Co_3O_4@NiO core@shell nanowire arrays for EES devices with high capacitance and high cycling stability because of the porous microstructural configuration and a balanced combination of two electrochemically active materials are reported.[47] Co_3O_4@NiO electrodes showed a high specific capacitance of 853 F×g^{-1} at a current rate of 2 A×g^{-1} after 6000 cycles. In another report, SnO_2 nanowire arrays coated with V_2O_5 were studied to bridge the performance gap between batteries and supercapacitors for high-rate LIBs.[48] An electrode made from the SnO_2/V_2O_5 core/shell nanowires demonstrate energy density of 282 W×h×kg^{-1} at a high power density of ~60 kW×kg^{-1}. A capacitance retention of ~67% is reported with a variation of charging-discharging current density from 0.2 to 20 A×g^{-1} up to 40 cycles.

A separate study used a nanotree forest of branched nanowire arrays to enhance electrochemical performance over 1D nanowire arrays, showing potential applications for energy storage devices at the nanoscale.[49] In a similar configuration, the 3D ZnO@MnO_2 core@shell nanostructures are reported to effectively increase the exposed area of electroactive materials to the electrolyte, loading quantity of electroactive materials per unit substrate area, and carrier transport in the hybrid electrode.[50] In a novel study, for the first time, the 3D ZnO@MnO_2 core@shell nanoforest electrodes showed five times higher capacitance (~31 $mF\times cm^{-2}$ at a current density of 20 $\mu A\times cm^{-2}$ for the nanoforest configuration), better rate capability, and longer charge-discharge stability at higher current densities, as compared with those for nanowire arrays. An 87% retention of capacitance is reported in the former compared with 26% for the latter, with a variation of charging-discharging current density from 0.02 to 0.2 $mA\times cm^{-2}$.

In a seminal study, hierarchical nanowire and atomically thin nanosheet (~2.0 nm) arrays of nickel cobalt oxide are reported with enhanced energy storage properties.[51] The structure generated abundant oxygen vacancies and antisite defects, and electrodes made out of the materials showed a more than fivefold increase in specific capacity, higher rate capability up to 100 $A\times g^{-1}$, and exceptional cycling stability of 10,000 cycles at 50 $A\times g^{-1}$ in alkaline electrolyte compared with those for pristine sample. The enhanced properties in the presence of defects are validated using density functional theory (DFT) simulations.

3.2.2 Photovoltaic Energy Storage

A quasi-solid state, coaxial dye-sensitized solar cell (DSC) fibre was reported by wrapping a transparent conducting layer of carbon nanotube on a photoactive dye-coated TiO_2 nanotube cladded Ti wire.[52] The TiO_2 nanotubes were grown by electrochemical anodization of Ti wire, and their growth orientation was normal on the wire surface. The highest energy conversion efficiency of 2.6% was measured for the dye-coated nanotube length of ~32 µm.[53] Similar fibrous dye-sensitized solar cells with TiO_2 nanotube array on Ti wire as the photoanode (twisted by a Pt wire counter electrode) have been reported. It was shown that the thread pitch significantly affected the photovoltaic performance of these fibrous solar cells. For an optimized thread pitch value of 1 mm for fibrous DSCs, conversion efficiency has been remarkably improved to 5.84%.[52]

3.2.3 Catalysis

TMOs are deliberated as cost-effective oxygen evolution reaction (OER) catalysts to replace precious noble metal-based catalysts. A series of Fe-doped $NiCo_2O_4$ (NCO) nanowires with different Fe-doped concentrations were studied as high-efficiency electrocatalysts for OER. These NCO nanowires demonstrate low overpotential and small Tafel slope in the Btler-Volmer expression because of ample catalytically active sites, high charge transport capability and specific surface area.[54] An outstanding OER performance with a low overpotential of 297 mV at a current density of 10 $mA\times cm^{-2}$ and a small Tafel slope of about 68 $mV\times dec^{-1}$ in 1.0 M KOH was reported for optimized Fe-doped sample demonstrating a promising electrocatalyst for the OER in water splitting.[54]

A visible light response of WO_3 nanofibers-C/Cu_2O nanowire array using the principle of dual-photoelectrode solar-charged photoelectrochemical wastewater fuel cell was reported for efficient hydrogen production based on the promotion of phenol oxidation at the anode.[55] A maximum amount of 93.08 $\mu mol\times cm^{-2}$ H with a total organic carbon removal rate of 82.12% was demonstrated under visible light irradiation for

8 h without additional bias. The value was 3.02 times higher than that of pure photocatalytic water splitting.

A microbial photoelectrochemical cell was reported with a *p*-type Co_3O_4 nanorod-arrayed photocathode for CO_2 conversion to formic acid with a yield of 239 ± 10 μmol in 10 h, and the maximum power density was 331 ± 4 $mW \times m^{-2}$ under visible light.[56] A novel nanostructured plasmonics Ag and AgCl NPs supported on chiral TiO_2 nanofiber-based plasmonic photoanode demonstrated treatment of urban wastewaters with simultaneous hydrogen production.[57]

The interfacial effect of a catalytic noble metal (Pt) with various transition metal oxides (TiO_2, CoO) on the heterogeneous catalytic activity was comprehensively explored.[58] In the study, the interfacial effect of cross-stacked Pt and TiO_2 or CoO nanowires possessing a strong metal-support interaction was reported to exhibit oxidation of CO. The array of Pt/TiO_2 nanowire and Pt/CoO nanowire catalysts demonstrated CO oxidation under 40 Torr CO and 100 Torr O_2 at 200–240°C with higher catalytic activity in the latter. The study deliberated the role of nanoscale metal-oxide interfaces in achieving novel catalytic materials.

3.3 GAS SENSING

Monitoring environmental quality and safety attracted immense attention looking at the modernization of life and agriculture, as well as industrial growth at its backdrop. The fouling species are generally known to be CO, NO, NO_2, H_2S, NH_3, HCHO, acetone, ethanol, and petroleum, along with elemental gas species, namely, H_2, O_2, which in excess can create either health or fire hazard. There are various methods of sensing these hazardous or fouling environmental species and industrial by-products. Among others, the chemiresistive technique is a vastly accepted sensing technology for semiconductor-based sensors encompassing TMOs of iron oxides, CuO, vanadium oxides, Co_3O_4, NiO, TiO_2, MoO_3, WO_3.[59,60] In this technique, gas sensing occurs at the semiconductor grain boundaries of the polycrystalline oxide film where reduction/oxidation processes take place. Other predominate processes of chemical reaction at the surface associated with pre-adsorbed ambient oxygen, physisorption of the chemical species directly on the semiconductor surface modifying surface states, and in general, complex surface chemical reactions between the different adsorbed chemical species, transfer of delocalized conduction electrons to localized surface states and vice versa. Catalytic effects also play a pivotal role in the chemiresistive sensing mechanism.[61-65]

3.3.1 Iron Oxides

In the TMO series, iron oxide-based sensors have generated interest for their cost-effectiveness, corrosion resistance, and high sensitivity. Among various polymorphs of iron oxides, the most stable hematite (α-Fe_2O_3) with rhombohedral corundum structure and metastable maghemite (γ-Fe_2O_3) with the cubic inverse-spinel structure are sensitive to the different gas environments of H_2, NO_2, volatile organic compounds (VOCs), and H_2S. The gas-sensing mechanisms in these two polymorphs are different because of their different crystal structures pertaining to the O occupancy sites. In the case of α-Fe_2O_3, Fe^{3+} occupy two-thirds of the octahedral sites in the hexagonal close-packed O lattice. In contrast, for γ-Fe_2O_3, one-sixth of Fe^{3+} ions at the octahedral sites are vacant to compensate for the charge, resulting in a defect (vacancy: o) containing inverse-spinel ($A(AB)O_4$:0.75[Fe^{3+}][$o_{1/3}Fe^{3+}{}_{5/3}$]O_4). Being exposed on the surface, these defects act as strong adsorption sites.[66] The surface-depletion mechanism is used to explain sensing in α-Fe_2O_3, as it is a surface-resistance-control material. On the other hand, being a bulk-resistance-control material, resistance change in γ-Fe_2O_3 arises from the phase formation between γ-Fe_2O_3 and magnetite (Fe_3O_4).[67] The thermal stability of γ-Fe_2O_3 owing to its phase transition to α-Fe_2O_3 in the temperature range of 523–873 K restricts the application of the former at elevated temperatures. Thus, α-Fe_2O_3 is the most used phase of iron (III) oxide for gas-sensing applications.

Along with phase, the morphology of sensing material plays a pivotal role in gas-sensing performance with a large effective surface area and an appropriate porous structure. The large area of nanostructures provides active sites for enhancing gas adsorption and improving sensitivity. Simultaneously, porous structure contributes toward fast response/recovery time because of the rapid diffusion of adsorbed species. In the past few decades, great efforts have been devoted to synthesizing various types of electrospun 1D α-Fe_2O_3, such as nanorods, nanotubes, and nanofibers.[68] La-, Nd-, Sm-, Ce-, and Ca-doped α-Fe_2O_3 nanotubes confirmed sensing acetone in the range of 500 ppb to 5 ppm at temperatures of ≤240°C with maximum response at <50 for 50-100 ppm .[69-73] Ca-doped and Al_2O_3-included nanotubes established 300 ppb to 5 ppm ethanol sensing in similar conditions.[74,75] Pt-functionalized α-Fe_2O_3 nanowire demonstrated H_2S sensing at 175 C with an optimized response of ~150 for 10 ppm.[76] On doping with SnO_2, α-Fe_2O_3 nanotubes sensed 1 ppm formaldehyde at 220°C with optimum response 25 for 500 ppm.[77] Composites of α-Fe_2O_3 with SnO_2 in the form of nanotube demonstrated 5 ppm sensing of both acetone and ethanol at 200°C with optimum response <25 for 100 ppm.[78] Similarly, hierarchically core-shell hollow nanofibers established 2 ppm of both ethanol and acetone sensing at an elevated temperature of 340°C with optimum response <30 at 100 ppm.[79] Cage-like nanofibers[78] and heterostructures nanotube[80] of α-Fe_2O_3/SnO_2 confirmed acetone sensing of 100 ppm at 275°C with response 5.3 for 100 ppm in the former and 1 ppm at 300°C with maximum response 33 for 100 ppm. In another report, In_2O_3-α-Fe_2O_3 composite nanotubes demonstrated 1 ppm of both formaldehyde and acetone sensing at ≤260°C with a response of <25 for 100 ppm.[81]

3.3.2 Copper Oxides

CuO is known to be a *p*-type metal oxide semiconductor and is exploited in various opto and electronic applications, including gas (H_2S, CO, NO_2, CO_2, and NH_3) sensing. However, in its pure form, *p*-type CuO is not very popular in sensing applications. Nevertheless, it is realized that CuO provides a good pathway for boosting sensing with the formation of homo- and heterojunction via loading of semiconducting or noble metal nanostructures. Nanopatterned channels of *p*-type CuO/Cu_2O functionalized with Ag nanoparticles established VOC sensing.[82] Thus, the role of noble metal activation was understood in prompting electronic and chemical sensitization effects along the sensing channel and enhancing the VOC gas-sensing performance. Composite nanofibers of CuO with SnO_2[83,84] and ZnO[85] showed a response to H_2S with a detection limit of 0.1-10 ppm in the temperature range of 15-300°C. Heterojunction nanofiber with SnO_2 showed sensing of CO with a response of 95 for 10 ppm.[86]

3.3.3 Vanadium Oxides

Among all vanadium oxide phases, vanadium pentoxide (V_2O_5) nanostructures are the widely studied phase for gas sensing because of their catalytic property along with their effective large surface at Van der Waals-bonded layered structure enabling adsorption of various analyte molecules.[87] The material has also been well studied for catalytic, thermo-, and electrochromic storage (utilizing 2D structure) applications, which are discussed in this book. Extremely high sensitivity (<30 ppb) for 1-butylamine and moderate sensitivity for ammonia was demonstrated for 1D V_2O_5 nanofibers.[88] The sensitivity to 1-butylamine increased linearly with the increase in the relative humidity of the reaction chamber by up to 60%. Configuration of contact formation (top/bottom) also affected the response. The sensing mechanism is deliberated, invoking intercalation of the amine into the layered structure of the nanofibers and sorption at the nanofibre/electrode interface. However, poor sensitivity was reported for toluene and 1-propanol vapours. Vanadium dioxide (VO_2) film with nanorod morphology, showing a reversible MIT at around 68°C, exhibited CH_4 sensing in a noble report (Figure 3.6).[89] The nanorods responded significantly to 50 ppm CH_4 in the insulating (semiconducting) phase at 50°C. Pd nanocluster functionalized single VO_2 nanowire demonstrated exceptional sensitivity to hydrogen originating from a large downward shift in the MIT subsequent to the adsorption and incorporation of atomic hydrogen.[90] Chemisorption of molecular H_2 on Pd activates the dissociation to atomic H, leading to a ~1000-fold increase in conductivity in the VO_2 nanowire. The transition temperature of MIT is decreased by >10°C instantly with a trace amount of H_2 exposure. Subsequently, H continues to diffuse into the VO_2 backbone for a significantly long time to saturate with only a modest change in the transition temperature. Change in the conductivity, however, was significant. Electron–phonon coupling in VO_2 modifying lattice normal modes with the introduction of H impurity and influence on electronic structure were the two prominent mechanisms involved in understanding the underlying physical process.

3.3.4 Cobalt Oxide

Co_3O_4, a simple spinnel comprising CoO and Co_2O_3 compositions, is another dominant *p*-type semiconductor used for gas sensing.[91] The catalytic action of *p*-type Co_3O_4 improves the selective oxidation of various VOCs, especially ethanol and acetone. Co_3O_4 nanowires and hollow carbon spheres showed significant sensitivity toward acetone at very low temperatures with optimum sensing at 150°C.[92] Electrospun nanofibers of Co_3O_4 demonstrated acetone sensing at 300°C with a response of 50 for 100 ppm.

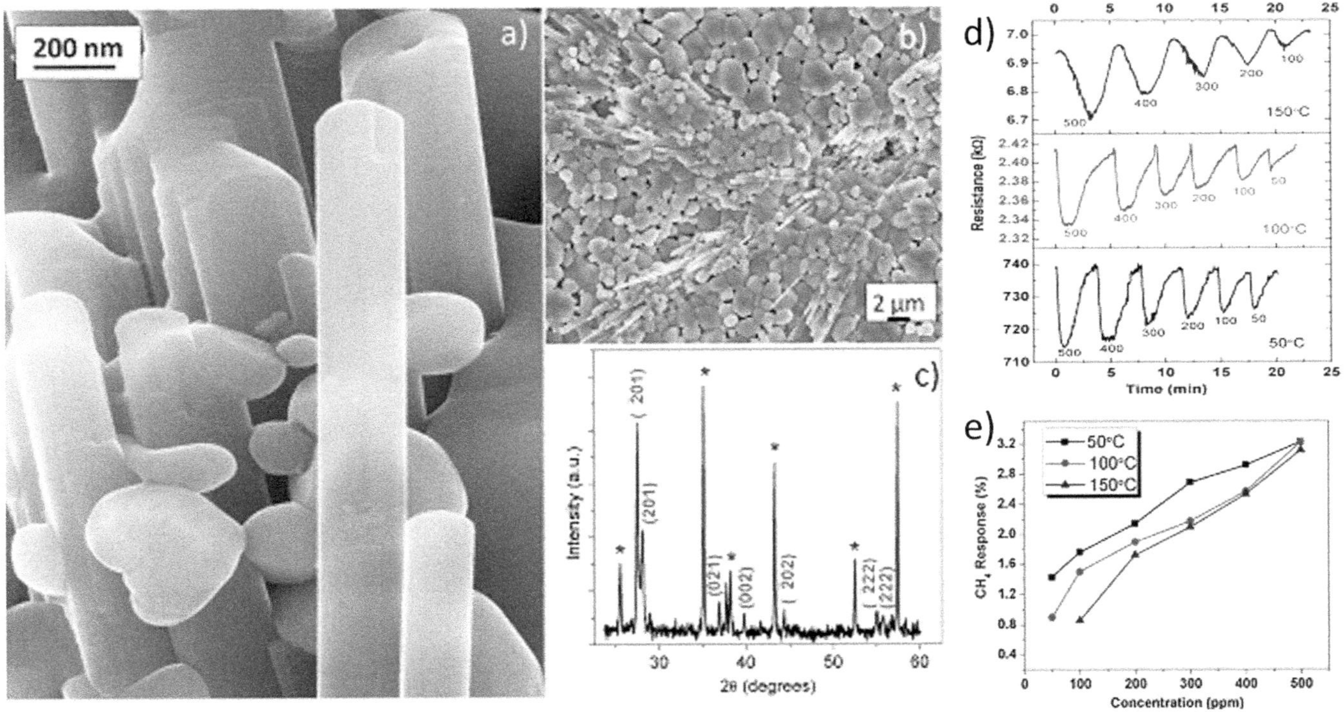

FIGURE 3.6 (**a**) High-magnification SEM image of VO_2 films, (**b**) low-magnification SEM image showing regions of nanorod bunches, (**c**) GIXRD pattern (* indicates substrate peaks). (**d**) Response transient curves of VO_2 toward CH_4 at 50°C, 100 C, and 150°C (numbers beneath *curves* indicate concentrations), (**e**) comparison of sensor response versus concentration at various operating temperatures. (Reproduced with permission from [89]. Copyright 2013, Elsevier B. V.)

3.3.5 Nickel Oxide

As discussed earlier, *p*-type TMOs have received special attention for the easy formation of semiconducting heterojunctions with most of the metal oxides, which are generally *n*-type. Chemo-conductivity shows enhanced sensitivity in the heterojunction configuration. NiO, a chemically stable wide bandgap (3.6-4.0 eV) *p*-type semiconductor with excellent electrical properties, is an important sensor material along with other applications in photovoltaics, fuel cells, and batteries.

Mesoporous NiO nanowire sensor demonstrated high sensitivity, fast response, and recovery to ethanol at 340°C as compared with those for the dispersed NiO nanowires. The higher surface area in the former is responsible for the reported enhancement.[93] Hydrothermally grown NiO nanowires established selectivity and sensitivity to 100 ppm of formaldehyde at 200°C.[94] Continuous architecture with hierarchical net-like NiO nanowire demonstrated acetone sensing of 200 ppm at 300°C and an improved response compared with its counterpart discontinuous net-like nanorods constructed NiO nanostructures.[95] The electrical pathway of charge carriers in the continuous network of the former significantly affects the sensitivity. Electrospun NiO nanofibers confirmed a faster recovery to H_2 and NH_3 compared with their bulk counterparts.[96] Using a similar growth technique, nanowebs of *p*-type NiO/*n*-type SnO_2 heterojunctions showed 100 ppm ethanol sensing at 300°C with a response value of 27.5.[97]

In a novel technique, self-powered gas sensing was expedited using piezoelectric NiO/ZnO heterojunction nanowire. H_2S sensing of 100 ppm at room temperature was reported with a moderate response value of 31.5 in the nanowire arrays, and NiO/ZnO nanogenerator was the power source.[98] At an exposure of 1000 ppm H_2S at room temperature, the piezo-voltage of the heterojunction assembly in dry air decreases to 61 mV from ~390 mV, and the response is one order higher than that of uncoated ZnO nanowire arrays. Hierarchical heterostructures of ZnO-NiO nanowires showed sensing of acetone, ethanol, NO_2, and H_2. The sensing mechanism is mainly dominated by *p-n* (*p*-NiO-*n*-ZnO) heterojunction with a moderate response of 95 and recovery times for 100 ppm of ethanol gas at a temperature of 500°C.[99]

3.3.6 Titanium Dioxides

Wide and predominantly indirect bandgap of ~3 eV has made TiO_2 an antiquated material for sensor applications in the past decades. It is an *n*-type semiconductor with anatase and rutile phases being used as the most important functional materials because of their corrosion resistance, high stability, and bio- and environment compatibility.

TiO_2 nanofibers showed sensitivity to CO around 200°C with a response around 3–4 for 25 ppm.[100,101] CO detection was reported even at room temperature using hierarchical nanofibers of TiO_2.[102] TiO_2 nanofibers also demonstrated NO_2 sensing with a response of around 30 for 50 ppm at 450°C.[103] TiO_2 nanorods showed acetone sensing at 500°C with a response around 9 for 10 ppm.[104] Ag- and Pd-functionalized TiO_2 nanofibers established H_2S (1 ppm at 350°C with a response of 120 for 1 ppm)[105] and NO_2 (0.16 ppm at 160°C with a response of 38 for 2 ppm)[106] sensing, respectively.

Nanocomposites of ZnO-TiO_2 hierarchical heterojunctions demonstrated ethanol sensing of 20 ppm at 320°C with a response of 50 for 500 ppm.[107] Similarly, SnO_2-ZnO core-shell heterostructure nanofibers confirmed sensing 10 ppm acetone at 280°C with a response of 13 for 100 ppm.[108] In_2O_3 Beads@TiO_2–In_2O_3 and In_2O_3/TiO_2 nanofibers reported 0.3 ppm and 97 ppb NO_2 sensing at room temperature with a response of 95 and 41 for 97 ppm, respectively.[109,110] The pristine and NiO nanoparticle decorated TiO_2 (NiO@TiO_2) nanorod sensors showed responses of ~6 and ~9, respectively, for 200 ppm acetone at 400°C. The latter also showed a short sensing time than the latter. The enhanced sensing performance of the NiO@TiO_2 nanorod sensor is predominantly because of modulation of the potential barrier height at the interface, high catalytic activity of NiO, and creation of active adsorption sites on NiO nanoparticles.[111]

3.3.7 Molybdenum Trioxides

MoO_3 is an indirect bandgap (2.7–3.2 eV) *n*-type semiconductor with two prime polymorphs of stable orthorhombic α-MoO_3 with layered structure and metastable β-MoO_3 having a ReO_3-like structure. There can be a phase transition from β- to α-phase by glide along crystallographic shear planes. Like iron oxide, stable α-phase is largely used for sensing applications in MoO_3.[112]

Both NO_2 and $N(CH_3)_3$ (trimethylamine, TMA) are sensed by α-MoO_3 nanorods. A very high response of 542 is reported for NO_2 sensing with a detection limit in the range of 10–500 ppm at 290°C.[113] TMA was detected in the range of 0.1-100 ppm at 200-350°C with a response value of ~150.[114] Chemically grown MoO_3-SnO_2 nanofibers accomplished 500–300 ppm CO sensing at 200-400°C.[115] Most recently, a two-step hydrothermally grown 1D MoO_3@SnO_2 core-shell heterostructure exhibited 100 ppm ethanol sensing at an optimum operating temperature of 200°C.[116] A response of ~49 was reported for the heterostructure, and the value is one order higher than that measured for the pure MoO_3 phase and ~8 times higher than that of SnO_2. For a specific composition of Mo:Sn = 9:10 in the heterostructure, a fast response recovery, good selectivity, and excellent stability were also claimed among all composites.

3.3.8 Tungsten Trioxides

WO_3 is a wide bandgap *n*-type gas-sensing semiconductor. It is generally exploited at gas sensors focusing on environment protection, healthcare, disease diagnosis, and food safety.[117] Chimiresitors based on WO_3 sensors have advantages over micro-electromechanical systems (MEMS) sensors, optical sensors, surface acoustic sensors, and electrochemical sensors. As generally understood, the chemical reaction of the analyte is strongly influenced by the microstructures and surface states of sensing materials. Thus, the morphology and microstructure of WO_3 with specific active sites and gas diffusion pathways influence the sensing process.

Electrospun pristine, and La_2O_3-, Cu-, and Pt-activated WO_3 nanofibers demonstrated sensing acetone. In the pristine nanotube, sensing is performed at an optimum temperature of 270°C with a response of 55 for 50 ppm analyte.[118] In the functionalized samples, however, sub-ppm detection of acetone was reported at 350°C.[119-121] WO_3 nanotube and In_2O_3-WO_3 heterostructure nanofibers also exhibited acetone sensing with response values of 20 for 40 ppm analyte at 250°C in the former,[122] and 13 for 50 ppm at 275°C with a detection limit of 1 ppm in the latter.[123] Pristine WO_3 nanofiber is also used for sensing NO_2 with a response of 12 for 400 ppb analyte at 75°C,[124] and NH_3 having a response of 5.5 for 100 ppm analyte at 200°C.[125] Functionalization with Pd demonstrated 1 ppm H_2S sensing at 350°C with a very low response of <2 for 1 ppm analyte, and also detection of 20 ppb toluene at 350°C with a response of 5.5 for 1 ppm analyte.[126] Au-functionalized nanofiber showed 1 ppm *n*-butanol sensing at 250°C with a response of 230 for 100 ppm analyte.[127]

REFERENCES

1. G. Gruner, *Density Waves in Solids*, Addison-Wesley, Reading, MA (1994).
2. G. Travaglini, P. Wachter, *Solid State Commun.* 37 (1981) 599.
3. R. E. Peierls, *Quantum Theory of Solids*, Clarendon, Oxford (1955) 108.
4. W. Fogle, J. H. Perlstem, *Phys. Rev. B* 61 (1972) 1412.
5. H. J. Schulz, *J. Phys. France* 50 (1989) 2833.
6. D. Poilblanc, T. M. Rice, *Phys. Rev. B* 39 (1989) 9749.
7. J. Zaanen, O. Gunnarson, *Phys. Rev. B* 40 (1989) 7391.
8. A. R. Moodenbaugh, Y. Xu, M. Suenaga, T. J. Folkerts, R. N. Shelton, *Phys. Rev. B* 38 (1988) 4596.
9. M. K. Crawford, R. L. Harlow, E. M. McCarron, W. E. Farneth, J. D. Axe, H. Chou, Q. Huang, *Phys. Rev. B* 44 (1991) 7749.
10. Y. Nakamum, S. Uchida, *Phys. Rev. B* 48 (1992) 5841.
11. S. A. Kivelson, V. J. Emery, Coexistence of, and Competition between, Superconductivity, *Strongly Correlated Electronic Materials: The Los Alamos Symp. 1993* (*ed.* K. S. Bedell) Addison-Wesley, Reading, MA (1994) 619.
12. J. M. Tranquada, D. J. Buttrey, V. Sachan, J. E. Lorenzo, *Phys. Rev. Lett.* 73 (1994) 1003.
13. V. Sachan, D. J. Buttrey, J. M. Tranquada, J. E. Lorenzo, G. Shimne, *Phys. Rev. B* 51 (1995) 12742.
14. J. M. Tranquada, J. E. Lorenzo, D. J. Buttrey, V. Sachan, *Phys. Rev. B* 52 (1995) 3581.
15. J. M. Tranquada, B. J. Sternlleb, J. D. Axe, Y. Nakamura, S. Uchlda, *Nature* 375 (1995) 561.
16. M. C. Boyer, W. D. Wise, K. Chatterjee, M. Yi, T. Kondo, T. Takeuchi, H. Ikuta, E. W. Hudson, *Nat. Phys.* 3 (2007) 802.
17. C. Howald, H. Eisaki, N. Kaneko, M. Greven, A. Kapitulnik, *Phys. Rev. B* 67 (2003) 014533.
18. M. Vershinin, S. Misra, S. Ono, Y. Abe, Y. Ando, A. Yazdani, *Science* 303 (2004) 1995.
19. T. Hanaguri, C. Lupien, Y. Kohsaka, D.-H. Lee, M. Azuma, M. Takano, H. Takagi, J. C. Davis, *Nature* 430 (2004) 1001.
20. W. D. Wise, M. C. Boyer, K. Chatterjee, T. Kondo, T. Takeuchi, H. Ikuta, Y. Wang, E. W. Hudson, *Nat. Phys.* 4 (2008) 696.
21. R. Lindsay, W. Strange, B. L. Chamberland, R. O. Moyer Jr., *Solid State Commun.* 86 (1993) 759.
22. G. Cao, S. McCall, J. E. Crow, R. P. Guertin, *Phys. Rev. Lett.* 78 (1997) 93.

23. G. Cao, V. Dobrosavljevic, S. McCall, J. E. Crow, R. P. Guertin, *Physica B* 259-261 (1999) 951.
24. G. Cao, J. Bolivar, S. McCall, J. E. Crow, R. P. Guertin, *Phys. Rev. B* 57 (1998) 11039R.
25. H.-R. Ott, *Crystalline Electric Field and Structural Effects in f-Electron Systems*, Plenum Press, New York (1980).
26. V. N. Kostur, P. B. Allen, *Phys. Rev. B* 56 (1997) 3105.
27. P. Liu, K. Zhu, Y. Gao, H. Luo, L. Lu, *Adv. Energy Mater.* 7 (2017) 1700547 and references there in.
28. X. Lai, J. E. Halpert, D. Wang, *Energy Environ. Sci.* 5 (2012) 5604.
29. J. Wang, H. Tang, L. Zhang, H. Ren, R. Yu, Q. Jin, J. Qi, D. Mao, M. Yang, Y. Wang, P. Liu, Y. Zhang, Y. Wen, L. Gu, G. Ma, Z. Su, Z. Tang, H. Zhao, D. Wang, *Nat. Energy* 1 (2016) 16050.
30. P. Liu, Y. Xu, K. Zhu, K. Bian, J. Wang, X. Sun, Y. Gao, H. Luo, L. Lu, J. Liu, *J. Mater. Chem. A* 5 (2017) 8307.
31. C. Niu, J. Meng, C. Han, K. Zhao, M. Yan, L. Mai, *Nano Lett.* 14 (2014) 2873.
32. T. Zhai, H. Liu, H. Li, X. Fang, M. Liao, L. Li, H. Zhou, Y. Koide, Y. Bando, D. Golberg, *Adv. Mater.* 22 (2010) 2547.
33. L. Mai, F. Dong, X. Xu, Y. Luo, Q. An, Y. Zhao, J. Pan, J. Yang, *Nano Lett.* 13 (2013) 740.
34. H. Liu, W. Yang, *Energy Environ. Sci.* 4 (2011) 4000.
35. L. Li, Z. Wu, S. Yuan, X.-B. Zhang, *Energy Environ. Sci.* 7 (2014) 2101.
36. H. Gwon, J. Hong, H. Kim, D.-H. Seo, S. Jeon, K. Kang, *Energy Environ. Sci.* 7 (2014) 538.
37. L. Wen, F. Li, H. M. Cheng, *Adv. Mater.* 28 (2016) 4306.
38. S. Wan, J. Peng, L. Jiang, Q. Cheng, *Adv. Mater.* 28 (2016) 7862.
39. J. W. Lee, S. Y. Lim, H. M. Jeong, T. H. Hwang, J. K. Kang, J. W. Choi, *Energy Environ. Sci.* 5 (2012) 9889.
40. C. Niu, J. Meng, C. Han, K. Zhao, M. Yan, L. Mai, *Nano Lett.* 14 (2014) 2873.
41. M. S. Balogun, Y. Luo, F. Lyu, F. Wang, H. Yang, H. Li, C. Liang, M. Huang, Y. Huang, Y. Tong, *ACS Appl. Mater. Interfaces* 8 (2016) 9733.
42. P. Sivakumar, M. G. Jung, C. J. Raj, J.-W. Park, H.-S. Park, H. Jung, *Int. J. Energy Res.* 45 (2021) 1809.
43. D. Zhao, M. Dai, Y. Tong, X. Song, X. Wu, *CrystEngComm* 21 (2019) 5789.
44. X. Xia, J. Tu, Y. Zhang, X. Wang, C. Gu, X.-B. Zhao, H. J. Fan, *ACS Nano* 6 (2012) 5531.
45. J. Zhou, D. Li, J. Han, X. Fan, *Dalton Trans.* 48 (2019) 2741.
46. J. P. Liu, J. Jiang, C. W. Cheng, H. X. Li, J. X. Zhang, H. Gong, H. J. Fan, *Adv. Mater.* 23 (2011) 2076.
47. X. H. Xia, J. P. Tu, Y. Q. Zhang, X. L. Wang, C. D. Gu, X. B. Zhao, H. J. Fan, *ACS Nano* 6 (2012) 5531.
48. J. Yan, A. Sumboja, E. Khoo, P. S. Lee, *Adv. Matter.* 23 (2011) 746.
49. S. H. Ko, D. Lee, H. W. Kang, K. H. Nam, J. Y. Yeo, S. J. Hong, C. P. Grigoropoulos, H. J. Sung, *Nano Lett.* 11 (2011) 666.
50. Q. Li, *Nanostructured transition metal oxides for energy storage and conversion* (2014). Theses and dissertations – UTB/UTPA. 920. https://scholarworks.utrgv.edu/leg_etd/920.
51. T. Wang, B. Xu, Y. Wang, J. Lei, W. Qin, K. Gui, C. Ouyang, K.-J. Chen, H. Wang, *Chin. Chem. Lett.* 33 (2022) 2669.

52. S. Huang, X. Guo, X. Huang, Q. Zhang, H. Sun, D. Li, Y. Luo, Q. Meng, *Nanotechnology* 22 (2011) 315402.
53. H. Sun, H. Li, X. You, Z. Yang, J. Deng, L. Qiu, H. Peng, *J. Mater. Chem. A.* 2 (2013) 345.
54. Z. Hao, P. Wei, H. Kang, Y. Yang, J. Li, X. Chen, D. Guo, L. Liu, *Nanotechnology* 31 (2020) 435707.
55. Z. Zhou, Z. Wu, Q. Xu, G. Zhao, *J. Mater. Chem. A Mater.* 5 (2017) 25450.
56. J. Wu, X. Han, D. Li, B. E. Logan, J. Liu, Z. Zhang, Y. Feng, *Appl. Catal. B Environ.* 276 (2020) 119102.
57. D. Wang, Y. Li, G. L. Puma, C.Wang, P. Wang, W. Zhang, Q. Wang, *Appl. Catal. B Environ.* 168-169 (2015) 25.
58. H. C. Song, G. R. Lee, K. Jeon, H. Lee, S. W. Lee, Y. S. Jung, J. Y. Park, *ACS Nano* 14 (2020) 8335.
59. H. Wang J. Ma, J. Zhang, Y. Feng, M. Teja Vijjapu, S. Yuvaraja, S. G. Surya, K. N. Salam, C. Dong, Y. Wang *J. Phys. Condens. Matter.* 33 (2021) 303001.
60. G. Korotcenkov, *Matter. Sci. and Engg. B* 139 (2007) 1.
61. S. Parida, A. Das, A. K. Prasad, J. Ghatak, S. Dhara, *Phys. Chem. Chem. Phys.* 20 (2018) 18391.
62. A. Patsha, P. Sahoo, S. Amirthapandian, A. K. Prasad, A. Das, A. K. Tyagi, M. A. Cotta, S. Dhara, *J. Phys. Chem. C* 119 (2015) 21251.
63. B. Venkataramana, A. Das, N. G. Krishna, A. K Prasad, S. Dhara, A. K.Tyagi, *Appl. Phys. Lett.* 105 (2014) 243102.
64. A. Das, B. Venkataramana, A. K. Prasad, D. Panda, S. Dhara, A. K. Tyagi, *J. Mater. Chem. C* 2 (2014) 164.
65. P. Sahoo, S. Dhara, S. Dash, S. Amirthapandian, A. K. Prasad, A. K. Tyagi, *Intl. J. Hydrogen Energy* 38 (2013) 3513.
66. S.-S. Li, W.-J. Li, T.-J. Jiang, Z.-G. Liu, X. Chen, H.-P. Cong, J.-H. Liu, Y.-Y. Huang, L.-N. Li, X.-J. Huang, *Anal. Chem.* 88 (2016) 906.
67. M. Wang, T. Hou, Z. Shen, X. Zhao, H. Ji, *Sens. Actuators B* 292 (2019) 171.
68. M. Imran, N. Motta, M. Shafiei, *Beilstein J. Nanotechnol.* 9 (2018) 2128.
69. A. Katoch, G.-J. Sun, S.-W. Choi, J.-H. Byun, S. S. Kim, *Sens. Actuators B* 185 (2013) 411.
70. A. Katoch, Z. U. Abideen, J.-H. Kim, S. S. Kim, *Sens. Actuators B* 232 (2016) 698.
71. S. Cho, D.-H. Kim, B.-S. Lee, J. Jung, W.-R. Yu, S.-H. Hong, S. Lee, *Sens. Actuators B* 162 (2012) 300.
72. X. Yu, F. Song, B. Zhai, C. Zheng, Y. Wang, *Phys. E* 52 (2013) 92.
73. A. Katoch, S.-W. Choi, H. W. Kim, S. S. Kim, *J. Hazard. Mater.* 286 (2015) 229.
74. M. Shafiei, F. Hoshyargar, J. Lipton-Duffin, C. Piloto, N. Motta, A. P. J. O'Mullane, *Phys. Chem. C* 119 (2015) 22208.
75. A. Katoch, S.-W. Choi, H. W. Kim, S. S. Kim, *J. Hazard. Mater.* 286 (2015) 229.
76. C. Jiang, G. Zhang, Y. Wu, L. Li, K. Shi, *CrystEngComm* 14 (2012) 2739.
77. W. Zheng, X. Lu, W. Wang, Z. Li, H. Zhang, Z. Wang, X. Xu, S. Li, C. J. Wang, *Colloid Interface Sci.* 338 (2009) 366.
78. S. Putthanarat, R. Eby, W. Kataphinan, S. Jones, R. Naik, D. Reneker, B. Farmer, *Polymer* 47 (2006) 5630.
79. L. Liu, T. Zhang, L. Wang, S. Li, *Mater. Lett.* 63 (2009) 2041.
80. W.-T. Koo, S.-J. Choi, N.-H. Kim, J.-S. Jang, I.-D. Kim, *Sens. Actuators B* 223 (2016) 301.

81. J. Kukkola, M. Mohl, A.-R. Leino, J. Mäklin, N. Halonen, A. Shchukarev, Z. Konya, H. Jantunen, K. Kordas, *Sens. Actuators B* 186 (2013) 90.
82. Y. M. Choi, S.-Y. Cho, D. Jang, H.-J. Koh, J. Choi, C.-H. Kim, H.-T. Jung, *Adv. Funct. Mater.* 29 (2019) 1808319.
83. B. Ding, M. Wang, X. Wang, J. Yu, G. Sun, *Mater. Today* 13 (2010) 16.
84. L. Liu, S. Li, J. Zhuang, L. Wang, J. Zhang, H. Li, Z. Liu, Y. Han, X. Jiang, P. Zhang, *Sens. Actuators B* 155 (2011) 782.
85. M. Zhao, X. Wang, J. Cheng, L. Zhang, J. Jia, X. Li, *Curr. Appl. Phys.* 13 (2013) 403.
86. B. Huang, C. Zhao, M. Zhang, Z. Zhang, E. Xie, J. Zhou, W. Han, *Appl. Surf. Sci.* 349 (2015) 615.
87. A. A. Akande, T. Mosuang, C. N. M. Ouma, E. M. Benecha, T. Tesfamichael, K. Roro, A. G. J. Machatine, B. W. Mwakikunga, *J. Alloys Compd.* 821 (2020) 153565.
88. I. Raible, M. Burghard, U. Schlecht, A. Yasuda, T. Vossmeyer, *Sens. Actuators B* 106 (2005) 730.
89. A. K. Prasad, S. Amirthapandian, S. K. Nori, S. Dhara, S. Dash, N. Murali, A. K. Tyagi, *Sens. Actuators B* 191 (2014) 252.
90. J. M. Baik, M. H. Kim, C. Larson, C. T. Yavuz, G. D. Stucky, A. M. Wodtke, M. Moskovits, *Nano Lett.* 9 (2009) 3980.
91. S.-J. Choi, L. Persano, A. Camposeo, J.-S. Jang, W.-T. Koo, S.-J. Kim, H.-J. Cho, I.-D. Kim, D. Pisignano, *Macromol. Mater. Eng.* 302 (2017) 1600569.
92. C. Zhang, L. Li, L. Hou, *Sens. Actuators B* 291 (2019) 130.
93. X. Li, D. Li, J. Xu, H. Jin, D. Jin, X. Peng, B. Hong, J. Li, Y. Yang, H. Ge, X. Wang, *Mater. Res. Bull.* 89 (2017) 280.
94. Z. Li, *Vacuum* 143 (2017) 50.
95. T. P. Mokoena, H. C. Swart, D. E. Motaung, *J. Alloys Compd.* 805 (2019) 267.
96. A. Khalil, J. J. Kim, H. L. Tuller, G. C. Rutledge, R. Hashaikeh, *Sens. Actuators B* 227 (2015) 54.
97. Y. Wang, H. Zhang, X. Sun, *Appl. Surf. Sci.* 389 (2016) 514.
98. Z. Qu, Y. Fu, B. Yu, P. Deng, L. Xing, X. Xue, *Sens. Actuator. B* 222 (2016) 78.
99. C. Baratto, R. Kumar, E. Comini, M. Ferroni, M. Campanini, *Nanotechnology* 28 (2017) 465502.
100. J. Sungpanich, T. Thongtem, S. Thongtem, *Mater. Lett.* 65 (2011) 3000.
101. L. Giancaterini, S. M. Emamjomeh, A. De Marcellis, E. Palange, C. Cantalini, *Proc. Eng.* 120 (2015) 791.
102. X. Bai, H. Ji, P. Gao, Y. Zhang, X. Sun, *Sens. Actuators B* 193 (2014) 100.
103. R. Arsat, M. Breedon, M. Shafiei, P. G. Spizziri, S. Gilje, R. B. Kaner, K. Kalantar-zadeh, W. Wlodarski, *Chem. Phys. Lett.* 467 (2009) 344.
104. S. Wei, G. Zhao, W. Du, Q. Tian, *Vacuum* 124 (2016) 32.
105. Y.-E. Miao, S. He, Y. Zhong, Z. Yang, W. W. Tjiu, T. Liu, *Electrochim. Acta* 99 (2013) 117.
106. D. D. Trung, N. D. Hoa, P. V. Tong, N. V. Duy, T. D. Dao, H. V. Chung, T. Nagao, N. V. Hieu, *J. Hazard. Mater.* 265 (2014) 124.
107. T. M. Righi, R. S. Almeida, M. A. d'Ávila, *Proc. Macromol. Symp.* (2012) 230.
108. J.-S. Jang, S.-J. Kim, S.-J. Choi, N.-H. Kim, M. Hakim, A. Rothschild, I.-D. Kim, *Nanoscale* 7 (2015) 16417.
109. N. Tucker, J. J. Stanger, M. P. Staiger, H. Razzaq, K. Hofman, *J. Eng. Fibers Fabr.* 7 (2012) 63.

110. A. H. Khoshaman, P. C. Li, N. Merbouh, B. Bahreyni, *Sens. Actuators B* 161 (2012) 954.
111. G.-J. Sun, H. Kheel, S. Park, S. Lee, S. E. Park, C. Lee, *Ceram. Int.* 42 (2016) 1063.
112. A. K. Prasad, *Study of gas specificity in molybdenum trioxide/tungsten oxide thin film sensors and their arrays.* PhD Thesis, State University of New York at Stony Brook, ProQuest Dissertations Publishing 3179613 (2005).
113. S. Bai, S. Chen, L. Chen, K. Zhang, R. Luo, D. Li, C. C. Liu, *Sens. Actuators B* 174 (2012) 51.
114. S. He, W. Li, L. Feng, W. Yang, *J. Alloys Compd.* 783 (2019) 574.
115. R. Nadimicherla, H.-Y. Li, K. Tian, X, Guo, *Solid State Ion.* 300 (2017) 128.
116. X. Wang, Y. Gao, Q. Zhang, X. He, X. Wang, *Sens. Actuators B* 382 (2023) 133484.
117. C. Dong, R. Zhao, L. Yao, Y. Ran, X. Zhang, Y. Wang, *J. Alloys Compd.* 820 (2019) 153194.
118. C. Feng, C. Wang, P. Cheng, X. Li, B. Wang, Y. Guan, J. Ma, H. Zhang, Y. Sun, P. Sun, J. Zheng, G. Lu, *Sens. Actuators B* 221 (2015) 434.
119. L. Xu, B. Dong, Y. Wang, X. Bai, Q. Liu, H. Song, *Sens. Actuators B* 147 (2010) 531.
120. W. Zheng, X. Lu, W. Wang, Z. Li, H. Zhang, Y. Wang, Z. Wang, C. Wang, *Sens. Actuators B* 142 (2009) 61.
121. X. Chi, C. Liu, J. Zhang, L. Liu, H. Li, Y. He, X. Bo, L. Liu, *J. Semicond.* 35 (2014) 064005.
122. W. Zheng, X. Lu, W. Wang, Z. Li, H. Zhang, Y. Wang, Z. Wang, C. Wang, L. Giancaterini, S. M. Emamjomeh, A. De Marcellis, E. Palange, A. Resmini, U. Anselmi-Tamburini, C. Cantalini, *Sens. Actuators*, 229 (2016) 387.
123. H. Chen, J. Di, N. Wang, H. Dong, J. Wu, Y. Zhao, J. Yu, L. Jiang, *Small* 7 (2011) 1779.
124. S. Wei, G. Zhao, W. Du, Q. Tian, *Vacuum* 124 (2016) 32.
125. L. Giancaterini, S. M. Emamjomeh, A. De Marcellis, E. Palange, A. Resmini, U. Anselmi-Tamburini, C. Cantalini, *Sens. Actuators B* 229 (2016) 387.
126. L. Xu, R. Xing, J. Song, W. Xu, H. Song, *J. Mater. Chem. C* 1 (2013) 2174.
127. M. Zhang, T. Ning, S. Zhang, Z. Li, Z. Yuan, Q. Cao, *Mater. Sci. Semicond. Process.* 17 (2014) 149.

4 Transition Metal Oxides at 0D

4.1 TMO QUANTUM DOTS

Size-dependent tunability of bandgap and efficient generation of charge carrier are the major characteristics of semiconductor quantum dots (QDs), which attract significant attention by making them attractive for diverse photocatalytic and electrocatalysis applications. Semiconducting QDs show outstanding physicochemical properties of increased effective area of exposure with high surface-to-volume ratio, enhanced absorption coefficients, and charge separation efficiency. Moreover, these QDs with significantly high effective area owing to increased surface-to-volume ratio are reported to possess unsaturated chemical bonds at the surface, leading to the depression of activation energy suitable for gas-sensing applications. Transition metal oxide (TMO)-based QDs are attractive for the abundance of TMO in nature. The inherent non-stoichiometry of oxygen in the compound leads to additional chemical activity, relatively high carrier lifetime, and stability. If not all, we will try to do a case study of the majority of these TMO QDs.

4.1.1 Photocatalysis

In recent times, environmental pollutants, namely organic dyes from the textile, coloured dye industry, and pharmaceuticals involved primarily in aquaculture and animal husbandry for bacterial treatments, have become a major concern.[1] Harmful carcinogenic complexes originating from the partial breakdown of dyes and antibiotics via various chemical degradation processes of hydrolysis, oxidation, and anaerobic decolourization may lead to catastrophic effects in a civilized society.[2] Thus, special attention is paid to effectively treating the organic dye using various methodologies. One of them is advanced oxidation processes (AOPs), which oxidize organic contaminants using short-lived strong oxidative radical species of hydroxyl radicals (•OH) and superoxides ($•O_2^-$).[3,4] Photocatalysis is one of the methodologies adopted in the AOPs,[5,6] where TMOs play an important role with photo-excited electrons and generated holes taking part in the oxidation process in degrading most of the organic compounds into H_2O and CO_2.[7] Photocatalysis assisted with TMOs has many advantages over the traditional wastewater purification methodologies: (a) the efficiency of photocatalytic degradation processes is exceedingly high, because of the

DOI: 10.1201/9781003623694-4

strong redox power; (b) the former is energy efficient using renewable solar emission; (c) photocatalysis is a green technique with most photocatalysts and its by-products being nontoxic inorganic metal oxides; and (d) photocatalysis is an economic process as the catalysts are recyclable.[8]

TiO_2 is known to be a thermally stable, low-cost, and nontoxic photocatalyst with high carrier lifetime. TiO_2 QDs synthesized using solgel process and with a surface area of 45.791 $m^2 \times g^{-1}$ showed outstanding catalytic degradation (95%) of Indigo Carmine dye under 60 min of UV irradiation for a loading of 1 $g \times L^{-1}$ at pH 2.[9] Ultra-fine anatase TiO_2 QDs prepared in a similar solgel technique with particle size in the range of 3–7 nm showed degradation of 97.1% methylene orange (MO) and 97.9% methylene blue (MB) with the exposure of 80 min of UV light illumination.[10] The trapping experiments demonstrated the active role of •OH radicals in the degradation process. In another report, anatase TiO_2 QDs were grown with the solgel method in the presence of methanol (average particle size ~3 nm) and ethanol (average particle size ~7 nm).[11] A reduction in the bandgap to 2.85 eV (band gap of bulk anatase TiO_2 ~ 3.2-3.5 eV) of methanol-assisted QDs was reported because of the possible presence of surface -OH groups leading to higher catalytic degradation of MO than that for the QDs grown in ethanol-assisted process.

In a novel type-II *p-n* heterojunction photocatalysts of Cu_2O octadecahedron/TiO_2 QDs (average particle size ~4.7 nm) with a specific area of 11.2 $m^2 \times g^{-1}$ and bandgap ~2.1 eV showed 97% degradation of MO under 60 min visible light (Figure 4.1).[12] The heterojunction nanocomposites also showed superior stability, demonstrating their reusability for seven cycles. This superior catalytic activity was attributed to the migration of charge carriers, leading to increased carrier lifetime in the nano-particulate type-II heterojunctions and efficient light absorption at the interface.

Nanocomposites comprising graphitic (g)-C_3N_4 and TiO_2 QDs of 3-6 nm diameter in different proportions (0.25, 1, 3, and 5 wt%) formed by mechanochemical grinding technique showed a shift of band edge of the heterostructure toward the blue region as compared with pure g-C_3N_4.[13,14] From the electrochemical impedance spectroscopy analysis of the sample with 3 wt% TiO_2 QDs exhibited the highest amount of charge transfer as compared with the other samples indicating free migration of photo-excited charge carriers at the interfaces in the specific sample. A superior catalytic degradation of 78% toxic rhodamine blue (RhB) dye within 120 min of exposure to sunlight, with a rate constant ~3.4 times higher than that for pure g-C_3N_4, was reported for the sample with 3 wt% TiO_2 QDs. Holes and $\bullet O_2^-$ radicals are reported to be the active species for the degradation dye. The nanocomposite heterostructure showed excellent stability up to five repeated cycles. Increased carrier lifetime with the migration of charge carriers at the interface is responsible for the observed superior functionality. F-doped TiO_2 QD/g-C_3N_4 heterostructure was reported, showing photocatalysis mediated by chemical bonding, leading to efficient redox potential and the ability for interfacial charge transfer.[15] Nanocomposite TiO_2 QD/g-C_3N_4 type-II heterojunction established enhanced photocatalysis of 95.57% MO degradation efficiency with solar illumination, which is ~3.3 and 5.7 times higher than those of TiO_2 QDs and g-C_3N_4 nanostructures, respectively.[16] The type-II heterojunction activated

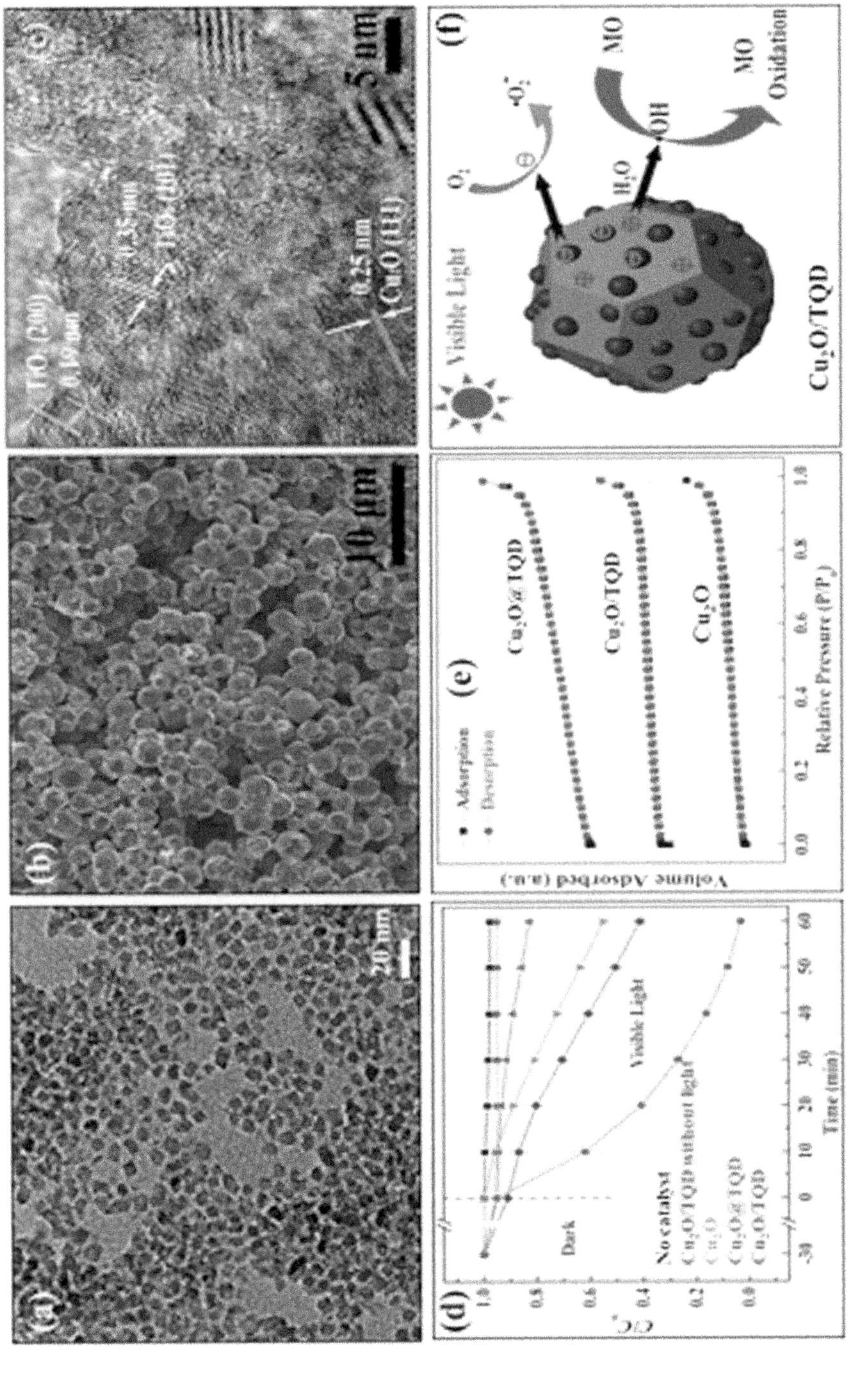

FIGURE 4.1 (**a**) TEM image of TiO_2 QDs (TQDs); (**b,c**) SEM and HRTEM image of Cu_2O/TQD nanocomposite; (**d**) photocatalytic performance of all prepared catalysts; (**e**) N_2 adsorption-desorption isotherms; and (**f**) proposed catalytic activity. (Reprinted with permission from Reference [12], Copyright 2016, American Chemical Society.)

effective photogenerated charge separation in the Z-scheme pathway, enabling photocatalysis under solar illumination.

The ternary system of anatase/rutile TiO_2 homojunction QDs on g-C_3N_4 nanosheets was reported for the degradation of antibiotic oxytetracycline in salt-water matrices.[17] The solar light adsorption occurs on the carbonaceous layers, while the photocatalysis continues on the g-C_3N_4 conduction band and anatase TiO_2 QDs valence band, generating oxidative free radicals for the degradation process with efficiencies approximately 3.8 and 46.8 times than that of TiO_2 and g-C_3N_4, respectively. The mechanism involved in the enhanced properties is due to enabled photogenerated electron-hole pairs and the combined effect between the adsorption and the catalysis processes.

The heterostructure of hydrothermally synthesized MoO_3 QDs on g-C_3N_4 nanosheets was reported to demonstrate photocatalysis for the reduction in U(VI) under solar illumination.[18] The 2% MoO_3 QD/g-C_3N_4 showed notable photocatalysis for the reduction process with effective separation of photogenerated carriers and the widened range of light absorption range.

Co_3O_4 QDs on P-doped g-C_3N_4 nanosheet grown using vitamin B_{12} were used for the photocatalytic degradation of metronidazole capitalizing effect of both P doping in g-C_3N_4 layer and Co_3O_4 QDs.[19] The type-II heterostructure enabled long-lived electron-hole pair generation and reduced bandgap energy for accelerating the photocatalysis.

Nanocomposites of CuO QDs/reduced graphene oxide (RGO) with varying loadings of RGO (0, 20, 50, 80, and 100 wt%) synthesized in a soft chemical route showed enhanced visible-light absorption and a reduction in the recombination rate in the type-II heterostructure with an enhanced lifetime of photogenerated charge carriers leading to the possible enhancement catalytic activity.[20] Photocatalysis study with 50 wt% CuO QDs/RGO nanocomposite heterostructure showed 99% degradation of MB within 60 min under visible-light illumination.

WO_3 QD-decorated GO with Mg-doped ZnO nanocomposites containing various loadings (0.16, 0.32, and 0.4 mg) of QDs grown using a co-precipitation technique was also reported for photocatalytic activity in RhB dye under sunlight with optimum result for samples having 0.32 wt% WO_3 QDs.[21] The formation of heterostructures was envisaged for the formation of enhancement of active sites at the surface with decreasing recombination of charge carriers in the process of efficient light absorption in the visible region.

4.1.2 Energy Storage

Reactive sputter-deposited transparent *p*-NiO QDs were studied for thickness-dependent electrochemical properties.[22] A Burstein-Moss (BM) blue shift was reported of bandgap from 3.6 to 3.9 eV, with enhancement in electrical conductivity with increased carrier concentration from 1.07×10^{20} to 2.24×10^{20} cm^{-3}, and higher optical absorption in the region of solar emission for thinner (~50 nm) sample. The degenerate NiO QDs showing the BM effect indicate a possible integration with low-bandgap materials forming heterojunction and utilizing increased absorption of light for the production of hydrogen via the water-splitting process.

$NiFe_2O_4$ QDs@Ni-based metalorganic framework (Ni-MOF) layered composite was reported as electrode material for supercapacitor (SC) application.[23] TMO QDs with multiple valence states and high electrical conductivity showed stability of MOFs and increased number of exposed active sites in the electrode, leading to the reduction in the diffusion path of electrolyte ions. Thus, fast charge transfer was possible with increased active redox reactions. A specific capacity of 852.3 $mA\times h\times g^{-1}$ at a current density of 1 $A\times g^{-1}$ was recorded for ten electrode $NiFe_2O_4$ QDs@Ni-MOF with an energy density of 32.5 $W\times h\times kg^{-1}$ at a power density of 798 $W\times kg^{-1}$ and a reasonable capacitance retention value of 87.3% at 1 $A\times g^{-1}$ up to 5000 cycles indicating potential applications of QDs and MOFs in the field of SCs.

Well-dispersed octylamine surfactant-coated spinel $MnFe_2O_4$ QDs on nanographene was reported as an active electrode for SC application with a high specific capacitance value of 517 and 905 $F\times g^{-2}$ for negative potential window −1 and − 1.2 V, respectively, in KOH electrolyte.[24] Condensed $MnFe_2O_4$ QDs@nanographene electrode with significant loading mass of 18 $mg\times cm^{-2}$ showed high areal (5.3 $F\times cm^{-2}$) and volumetric (277.6 $F\times cm^{-3}$) capacitances. A capacitance value of 150 $F\times g^{-1}$ at 200 $A\times g^{-1}$ was achieved and retained up to 65,000 cycles with the assistance of nanographene encapsulation in improving rate and cycling performance. Thus, QD-based electrodes may accelerate the development of energy storage device performance.

4.1.3 Gas Sensing

Anatase TiO_2 QDs decorated with 0-5 wt% Ag grown in the hydrothermal process were reported for ammonia (NH_3) sensing.[25] The sample with 3 wt% Ag demonstrated a high response (R_{air}/R_{gas}) value of 25.1 for 20 ppm NH_3 at room temperature (RT). The value reported is six times higher than undecorated TiO_2 QDs with a clear role of Ag as a catalyst.

Charge transfer-induced chemical enhancement in the surface-enhanced Raman spectroscopy (SERS)[26] effect has been utilized for $MoO_3{\cdot}xH_2O$ QDs (average diameter ~2.2 nm) to detect hydrazine with a minimum detection limit of 4×10^{-5} M (Figure 4.2).[27] A three-order enhancement as compared with those of nanoparticles of 10–100 nm was reported. The coupling of molecular vibration with defect states of MO_3 QDs is responsible for the observed SERS effect.

WO_3 QDs grown in a colloidal synthesis route were reported for H_2S gas detection with a response of 57 (maximum) for 50 ppm gas at 80°C.[28] A response time of 47 s and a recovery time of 126 s was also reported with a linear response in the 5–25 ppm detection range. A theoretical detection limit was interpolated in the study to a very low value of 56 ppb at 80°C. Monoclinic WO_3 QDs (average size ~4 nm) synthesized in the solvothermal technique showed sensing of 1-5 ppm NO_2 (oxidizing gas) at 100°C and 100–500 ppm C_2H_5OH (reducing gas) at 200°C.[29]

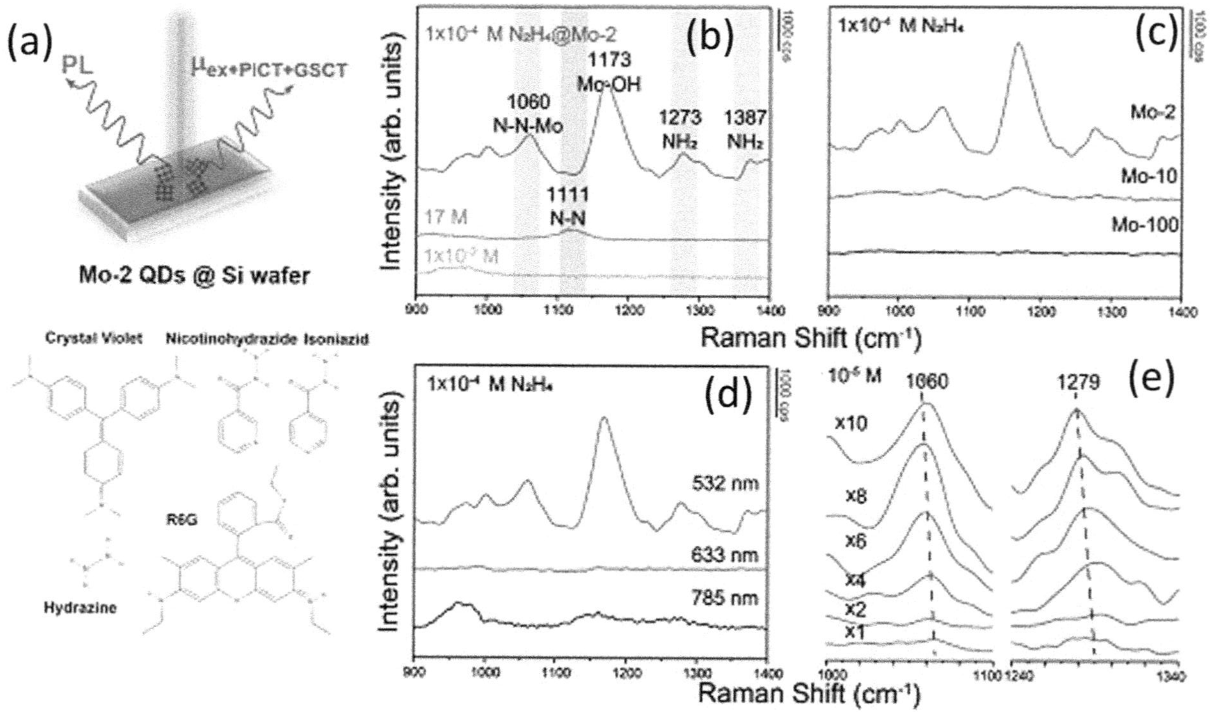

FIGURE 4.2 SERS performance evaluation of Mo-2, Mo-10, Mo-100 samples designated for ethanol:water volume ratios of 57.5:5, 52.5:10, and 42.5:20, respectively. (**a**) Schematic depiction of SERS platform fabricated from Mo-2 on Si substrate, and chemical structures of detectable probe molecules. (**b**) Typical Raman spectra of hydrazine solution (17 M, 10^{-2} M) and SERS spectra of N_2H_4 (10^{-4} M) collected on Mo-2. (**c**) Comparison of SERS spectra of N_2H_4 (10^{-4} M) collected on Mo-2, Mo-10, Mo-100. (**d**) Comparison of SERS spectra of N_2H_4 (10^{-4} M) collected on Mo-2 under 532, 633, and 785 nm laser excitation. (**e**) Peak profile of 1060 and 1279 cm^{-1} measured on Mo-2 at concentrations from top to bottom: 10×10^{-5}, 8×10^{-5}, 6×10^{-5}, 4×10^{-5}, 2×10^{-5}, 1×10^{-5} M. (Reprinted with permission from Reference [27], Copyright 2022, American Chemical Society.)

Presence of W^{5+} on the QD surface (reduced surface) was reported to oxidize at ambient conditions with slow kinetic (increased sorption of O_2), leading to the observed enhancement in sensing promoting increased adsorption of NO_2 and faster decomposition of C_2H_5OH.

4.2 TMO NANOCLUSTERS

There are plenty of studies on TMO nanoclusters for the applications discussed here. So, it will be unwise to discuss all the reports in the vast literature at the cost of various other interesting topics and properties discussed in this book. Instead, we would like to discuss the following applications of TMO nanoclusters in their different configurations.

4.2.1 PHOTOCATALYSIS

Heterojunction photocatalyst of TiO_2 nanocluster and 2D g-C_3N_4 (TiO_2/g-C_3N_4) grown by electrochemical etching and co-annealing method showed a unique degradation efficiency of tetracycline hydrochloride (TC) of about 99.04% in 120 min for solution pH value of 3–7 under Xenon lamp radiation.[30] In the photocatalytic degradation of the process of TC, the TiO_2/g-C_3N_4 heterojunction demonstrated exceptional anti-interference ability to NO_3^-, Cl^-, Na^+, Ca^{2+}, and Mg^{2+} ions in natural waters. Redox radicals of $\bullet O_2^-$ and •OH were substantiated as the foremost reactive species for the degradation process. In another study, 4 wt% g-C_3N_4/TiO_2 nanocluster (g-C_3N_4:TiO_2 = 1:25) demonstrated a reusable stable photocatalytic degradation efficiency of 96.53% for tetracycline hydrochloride under UV radiation.[31] Internal electric field on the step-scheme (S-scheme) heterojunction interface with the prime active species $\bullet O_2^-$, electron (e^-)-hole (h^+) pairs and •OH enhanced the photogenerated carries migration. The study established the application of S-scheme heterojunction photocatalysts for the photocatalytic degradation process. A heterojunction with 50 wt% TiO_2/high-crystalline g-C_3N_4 (HCCN) (TiO_2 nanoparticle dispersed on 2D HCCN) photocatalyst formed in a hydrothermal method was reported for the degradation efficiency of 90% for 10 mg×L^{-1} TC under visible ($\lambda > 420$ nm) radiation in 120 min with a reaction rate of approximately one order more than that of pristine HCCN and ~54 times better than that of pristine TiO_2.[32] The outstanding photocatalytic activity was accredited primarily to improved light absorption due to increased surface-to-volume ratio in the heterojunction and extended carrier lifetime with enhanced separation and migration of e^--h^+ pairs in the synergistic effect of the TiO_2/HCCN. The nanoparticle-2D heterojunction with 80 wt% g-C_3N_4/TiO_2 demonstrated optimum photocatalytic activity for the degradation of MB with reaction rates of 1.45, 2.20, and 1.24 times better than that of the pristine hydrothermally grown TiO_2 nanoparticle, g-C_3N_4 powder, and commercially available TiO_2 powder, respectively.[33] The improved efficiency was attributed to similar reasons as reported by Guo *et al.*[32] In another similar report, g-C_3N_4/TiO_2 with plasma-treated TiO_2 nanoparticle was reported to degrade TC utilizing •OH and $•O^{2^-}$ radicals originating from the plasma-treated TiO_2 nanoparticle-assisted dissociation of H_2O_2 generated by g-C_3N_4 photocatalyst.[34] Thus, an effective photo-Fenton degradation of TC was achieved in the TiO_2 (plasma treated)/g-C_3N_4 photocatalytic system.

Heterojunction of solvothermal grown Co_3O_4/g-C_3N_4 was reported to degrade atrazine up to 78.5% in 35 min with solar light illumination using $•SO_4^-$ radical formed in the synergistic catalysis process and was also validated using density functional theoretical calculations.[35] An inhibitory effect was found for CO_3^{2-} in the degradation process while in coexisting ions of NO_3^-, SO_4^{2-}, Cl^-, CO_3^{2-}. Whereas e^-, •OH, and $•SO4^-$ were the predominant active species in the gradation process.

In a hydrothermally grown Z-scheme heterojunction, MoO_3/g-C_3N_4 photocatalyst was reported for the degradation of TC.[36] Optimally grown 0.5 wt% MoO_3/g-C_3N_4 demonstrated removal efficiency of 85.9% in 100 min under visible light illumination with the rate constant of 2.3 times more than that of g-C_3N_4. It was found that $•O_2^-$ and h^+ were the prime active species in the stable photocatalytic process. In another report, MoO_3/g-C_3N_4 heterojunction-based photocatalyst was reported to remove RhB in a direct Z-scheme charge transfer process between MoO_3 and g-C_3N_4 with the accumulation of electrons in g-C_3N_4 and holes in MoO_3 and also assisted by the formation of $•O^{2^-}$ radicals.[37]

4.2.2 Energy Storage

Energy storage in the form of SCs has become popular these days with the quest for renewable green energy for environmental protection. SCs have many advantages over the conventional battery, namely, less weight and faster access with the same storage capacity, less charging time, and a charge/discharge cycle of ~10^6 times. SCs can store 10–100 times more energy than conventional capacitors.

As a major component of the SCs, TMO-based electrode materials are explored. The pseudocapacitive nature of the TMOs is due to the fast and reversible redox reactions at the surface of the electrode material.[38] MnO_2 nanoparticles are studied because of their natural abundance, thus cheap, and high theoretical capacitance value of 1370 $F•g^{-1}$. The charge storage mechanism is $MnO_2 + M^+ + e^- \leftrightarrow MnOOM$,[39] where M stands for electrolyte cations, for example, K^+, Li^+, Na^+. Charge transfer between multiple oxidation states of Mn^{2+} to Mn^{6+} contributes to the potential window and presence of various polymorphs of MnO_2, leading to its variation in electrochemical performance. However, the low conductivity of MnO_2 restricts its use as SC electrode material.[40] V_2O_5 is a Van der Waals-bonded compound where intercalation is possible between the layers.[41] The monovalent cation constitutes the oxide structure without changing the original structure of the material.[42] Layered V_2O_5 show superior potential for intercalation of various ions and can be used efficiently as SC as well as Li-ion batteries.[43] The charge storage mechanism is $V_2O_5 + 2M^+ + 2e^- \leftrightarrow M_2V_2O_5$. The hexagonal phase, among various other phases of WO_3, is reported to be showing the best performance of charge storage in the process of $WO_3 + M^+ + e^- \leftrightarrow HWO_3$.[44] Environment-friendly *p*-type NiO with low toxicity is reported to be an efficient electrode material with superior theoretical capacitance showing CV curves similar to battery materials. Observed redox peaks result from the mutual transformation of charge transfer between Ni^{2+} and Ni^{3+} which contributes to the recorded redox peak and diffusing OH^- species in the electrodes that take part in the charging and discharging process via $NiO + zOH^- \leftrightarrow zNiOOH + (1-z)NiO + ze^-$.[45]

In a novel approach, shortcomings of TMO electrodes are reduced by combining them with various other ions, namely, electrode material made by combining ZnO and NiO.[46] Specific capacitance was reported to improve with ZnO acting as the electrode modifier. In addition, NiO with multiple oxidation states helps the fast redox reaction, leading to the enhancement of specific capacitance. Similarly, SnO_2 was combined with NiO, exhibiting a high specific capacitance of 464 $F \times v \times g^{-1}$ at a scan rate of 5 $mV \times v \times s^{-1}$ along with a reasonable capacitance retention value of 87.24% after 1000 cycles.[47] In case of TMO-TMO heterostructures, namely, nanostructured MnO_2 and V_2O_5 core-shell is reported to have a high specific capacitance of 694 $F \times g^{-1}$ at a current density of 1 $A \times g^{-1}$.[48] Asymmetric SC using core-shell nanosheet of $ZnCo_2O_4$@MnO_2 showed a high specific capacitance of ~2170 $F \times g^{-1}$ at a current density value of 3 $mA \times cm^{-2}$ in KOH electrolyte.[49] Multicomponent properties of both the TMOs contribute to the stable performance. In the approach of device fabrication, TMO-activated C heterostructure with TMO and activated C as the positive and negative electrodes, respectively, showed an energy density of 29.41 $W \times h \times g^{-1}$ with 95.3% retention up to 3000 cycles.[49]

Cost-effective, abundant carbon-based materials with porous structure and adequate electrical conductivity are used as electrodes for SCs. Monoclinic VO_2 showed superior charge storage capacity with a Coulombic efficiency of 98.2% and a capacitance value of 33 $mF \times cm^{-2}$ having capacitance retention value of 93.7% up to 500 cycles with V^{+4} compared with VO_2 structures with mixed oxidation states of V^{4+} and V^{5+} of VO_2 (V^{5+}/V^{4+} increases progressively from sample S1 to S3) in triclinic and mixed phases (Figure 4.3).[50,51]

N-doped carbon@MnO_2 3D core-shell composite showed remarkable electrochemical capacitance of 247.9 $F \times g^{-1}$ in Na_2SO_4 electrolyte (1 M) at 0.5 $A \times g^{-1}$. A significant retention value of 82.9% after 2000 cycles at 5 $A \times g^{-1}$ was also recorded.[52]

Poor electrical conductivity of V_2O_5 and NiO can be secured by the proper combination with conducting materials, namely carbonaceous materials. Binder-free V_xO_y-embedded carbon fibre (CF) was reported as a SC device without the use of any conductive agent. Specific capacitance values of 557, 476, and 606 $F \times g^{-1}$ at 0.5 $A \times g^{-1}$ were reported for V_2O_3/CF, VO_2-V_2O_5/CF, and V_2O_5/CF configurations, respectively. The electrochemical performance of V_2O_5/CFs was recorded as superior to those of V_2O_3/CF and VO_2-V_2O_5/CF.[53] The core-shell hybrid NiO/RGO showed a specific capacitance of 940 $F \times g^{-1}$ for 2 $A \times g^{-1}$ current density.[54] In another report, hydrothermally grown NiO/RGO electrodes registered a specific capacitance value of 727.1 $F \times g^{-1}$ at a current density of 1 $mA \times cm^{-2}$ with a reasonable cyclic stability of about 80.4% after 9000 cycles.[55]

Two-dimensional materials have limitations in energy storage application, with moderate charge storage capacity being confined to the surface only, leading to a decrease in conductivity. High initial capacities of transition metal dichalcogenide

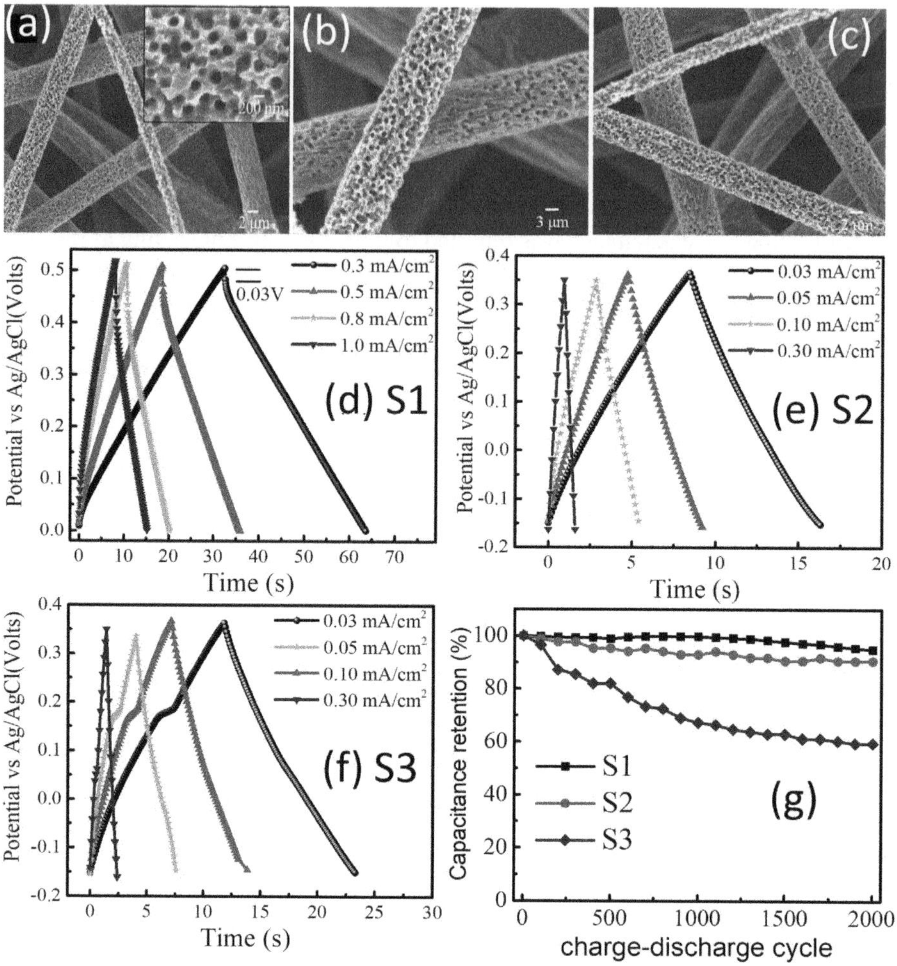

FIGURE 4.3 FESEM images of sample (**a**) S1, (**b**) S2, and (**c**) S3 prepared with Ar flow rates of 10, 20, and 30 sccm, respectively, with V^{5+}/V^{4+} increasing progressively from S1 to S3. Inset of (**a**) shows the magnified image with a typical pore diameter of approximately 200 nm. Charge-discharge profile at current density of 0.3-1.0 mA×cm^{-2} (**d**) S1, (**e**) S2, and (**f**) S3. (**g**) Capacitance retention of the samples with charge-discharge cycle. (Reprinted from Reference [50]. Creative Commons.)

(TMDs) are not very useful following a conversion reaction on the first discharge cycle, leading to poor retention capacity. However, few semiconducting TMDs have been reported to have adequate electronic conductivity.[56] Thus, studies of cost-effective TMD/TMO heterostructure are conducted for their SC performances. An asymmetric SC electrode device with a few-layered MoS_2-NiO/MoS_2-Fe_2O_3 configuration was reported, showing a maximum energy density and good cycling performance.[57] Hydrothermally grown 3D hierarchical nanosphere TiO_2-MoS_2 flake registered improved specific capacitance compared with pristine TiO_2.[58]

In another approach, pseudocapacitive conducting polymers (CPs) as electrode material were used to store charge at the surface and bulk of it. Poly(3,4-ethylenedioxythiophene) (PEDOT), polypyrrole (PPy), and polyaniline (PAni), and with their theoretical capacitance of about 100 $F \times g^{-1}$ are the prominent CPs. Two times higher specific capacitance was recorded for CPs than that of electric double-layer capacitors (EDLCs). In light of the low conductivity of TMOs, a combination of CPs and TMOs was envisaged to be advantageous for both in achieving high specific capacitance.[59] However, CPs show poor cyclic stability due to the variation in the polymer chain during the redox process.

4.2.3 Gas Sensing

Anatase TiO_2 nanoparticles showed ppb level detection of acetone at 270 C with fast response and recovery time. Enhanced adsorption activities of the TiO_2 surface are responsible for the selectivity in sensing acetone.[60] Anatase TiO_2 nanospheres grown on silica glass and plastic substrates were used for volatile organic compounds (VOCs), namely, methanol, ethanol, and 1-propanol.[61] The sensor responses decreased in the order of methanol < ethanol < 1-propanol at 50–350 ppm owing to the reductive nature of the gases depending on the number of CH chains. A very fast gas-sensing response (1 s) and recovery time (1 s) toward propylene glycol was recorded at 350°C owing to the large surface area and high crystallinity of nanospheres with a significant amount of 60% response for 50 ppm detection level. Laser-ablated TiO_2 nanoparticles were used to sense 200 ppm NH_3 with a remarkable 64% response at RT.[62] A moderated response (28 s) and recovery time (24 s) was achieved.

Noble metals nanocluster functionalized TiO_2 nanostructures are reported to show improved resistive gas-sensing performance owing to carrier transfer in the interface and subsequent electronic and chemical sensitization.[63] Pd-functionalized TiO_2 nanoparticles were reported as resistive gas sensors for butane.[64] A 7.5 mol% Pd-functionalized TiO_2 nanoparticles-based sensor showed a response value of 33.93 for 3000 ppm butane. The value is about one order higher than that of pure TiO_2 nanoparticles with a significant response (13 s) and recovery (8 s) time. $ZnFe_2O_4$ (ZFO) yolk-shell spheres made of ultrathin nanosheets and ultrasmall nanoparticles decorated with Au nanoclusters showed remarkable sensing of chlorobenzene with a fourfold increase in the response value of 90.9 as compared with undecorated ZFO and a low detection limit of 100 ppb for exposure of 10 ppm chlorobenzene at 150°C.[65] Density functional theory calculations showed that the enhanced sensing mechanism is primarily ascribed to the electronic and chemical sensitization prompted by the Au nanoclusters on the surface of ZFO.[65]

WO_3 decorated TiO_2 nanoparticles were studied for sensing xylene at different temperatures.[66] A significantly high sensor response of 92.53 for 10 ppm xylene at 160°C was reported for 10.0 mol% WO_3 decorated TiO_2 nanoparticles with a detection limit of 1 ppm.

The heterojunction of *p*-CuO nanosheet and *n*-TiO_2 nanoparticles was studied for detecting trace amounts of triethylamine (TEA). Using *p-n*-junction-assisted enhancement in sensing mechanism with carrier exchange at the heterojunction, *p*-CuO/*n*-TiO_2 sensors showed a high sensing response of 12.7 for 5 ppm TEA at 160°C with a low detection limit of 0.5 ppm.[67] In another study, the heterogeneous CuO decorated TiO_2

nanoparticle-based composites were reported for sensing NH_3 and H_2S. A 0.1 mol% CuO-decorated TiO_2 nanocomposites showed highest responses of 97 with a response time of 2 s to NH_3 and 77 with a response time of 45 s to H_2S for 50 ppm of each gas at RT.[68]

The sensing of NO_2 was reported to be close to RT using solvothermal grown *n*-WO_3 nanocrystals, forming a *p*-type layer on the adsorption of oxygen at lower temperatures.[69] Oxygen non-stoichiometric Co_3O_4-functionalized WO_3 hollow microspheres were reported for sensing toluene with a response value of 55.8 for 100 ppm of analyte.[70] Oxygen vacancies act as preferential sites for the chemisorption of toluene gas molecule, and Co_3O_4 contribute more toward it as the response of the composite was found to be almost three times higher than that of the undecorated WO_3 microstructures. A composite of TiO_2-WO_3 nanocrystallite was reported for NO_2 sensing utilizing a larger surface area achieved by forming finer particle sizes and agglomerates in the co-precipitation and precipitation processes than that achieved in the simple precipitation technique.[71]

A composite of porous Pd-WO_3 multilayer films was reported for 346.5 times more for hydrogen sensing than recorded with pure WO_3 thin film with a response time of 7 s for an optimal Pd:W molar ratio of 1%.[72] The Pd-WO_3 multilayer composite showed selectively higher ~20 times response to hydrogen than that for CO and CH_4. In another report, WO_3 nanoparticles grown by reacting sodium tungstate dehydrate and hydrochloric acid and subsequently calcining at 500 C showed significant response to CO sensing.[73] A homogenously dispersed 7.5 at% La-decorated WO_3 nanoparticles showed NO_2 sensing with a response value of 74.2 for 900 ppb at 125 C.[74] Low values of response (23 s) and recovery (35 s) times were also recorded. The sample also showed selectivity of NO_2 sensing in the presence of ppm-level NH_3 and SO_2 together as well as individual gases of NH_3, SO_2, CO, CO_2, and CH_4, along with tolerance to humidity fluctuations for stable performance. Ru (0.01–0.02 wt%) decorated WO_3 nanoparticles showed two order enhancements in sensor response for the detection of 0.5 ppm acetone.[75]

Fe-doped mesoporous WO_3 hollow nanospheres with a large effective surface area of ~165 $m^2 \times g^{-1}$ were reported for the selective detection of ppb-level NO_2 for air quality monitoring and asthma diagnosis by breath analysis.[76] Reduced cell parameters in doped WO_3 indicated distortion in the crystal lattice, inducing O non-stoichiometry suitable for gas sensing with possible adsorption of both O and NO_2 in the defect sites, enabling enhanced sensor performance. The 5.2% Fe-doped mesoporous WO_3 sensor showed remarkable sensitivity for detecting 10 ppb NO_2 at 120°C with a broad detection limit of 10–1000 ppb. C-doped WO_3 hollow nanospheres showed excellent acetone sensing with sensitivity down to 0.2 ppm with significant selectivity over ethanol, methanol, toluene, NH_3, NO, and CO.[77] The Si-doped WO_3 nanoparticles showed 50 ppb acetone detection with 50% increased sensitivity as compared with undoped WO_3.[78]

As compared with mesoporous WO_3 nanocrystals, an *n*-*n* homojunction of it with 5 wt% ZnO showed a significant selective sensing response of NO_2.[79] The homojunction-induced formation of depletion layer with separation of electron-hole pairs enabled charge transfer for the enhancement of sensing NO_2.

WO_3 nanosphere (diameter 200–400 nm) covered with graphene was reported for sensing NO_2 at RT with a response value of ~41 for an analyte concentration of 56 ppm.[80] The effective charge transfer taking place during the diffusion of the analyte through the large effect surface area of graphene across the composite interface was accountable for the observed sensing performance at RT. Composite of RGO and WO_3 was also reported for sensing NO_2 at RT.[81] Electron transfer processes in the conductive network in the increased surface area of RGO enabled RT detection of NO_2.

A PAni-WO_3 composite-based hybrid sensor showed NH_3 sensing in the protonation process of the analyte and the composite.[82] The composite sensor with 50 wt% WO_3 content demonstrated a very high response value of 121 for 100 ppm NH_3 at RT with selectivity in the presence of CO_2 and ethanol. A reasonable response value of 9 was also recorded for 1 ppm NH_3 at RT. On exposure to the analyte, the lone pair of electrons on the N atom of NH_3 donates holes (protons) to the N atoms of PAni and WO_3 nanoparticles, enhancing the protonation effect and influencing the sensing mechanism to a great extent. In another report, a composite of PPy and WO_3 nanoparticles was studied for sensing TEA with a significantly high response value of 680 for 100 ppm at RT.[83] Enhanced charge transfer and gas diffusion in *p*-PPy/*n*-WO_3 heterojunction were responsible for the enhanced sensitivity of the system.

REFERENCES

1. H. Anwer, A. Mahmood, J. Lee, K.-H. Kim, J.-W. Park, A. C. K. Yip, *Nano Res.* 12 (2019) 955.
2. X. Li, J. Xie, C. Jiang, J. Yu, P. Zhang, *Front. Environ. Sci. Eng.* 12 (2018) 14.
3. D. B. Miklos, C. Remy, M. Jekel, K.G. Linden, J. E. Drewes, U. Hübner, *Water Res.* 139 (2018) 118.
4. A. Tufail, W. E. Price, F. I. Hai, *Chemosphere* 260 (2020) 127460.
5. D. Zhu, Q. Zhou, *Environ. Nanotechnol. Monit. Manag.* 12 (2019) 100255.
6. H. Hou, X. Zhang, *Chem. Eng. J.* 395 (2020) 125030.
7. L. Zhang, J. Ran, S.-Z. Qiao, M. Jaroniec, *Chem. Soc. Rev.* 48 (2019) 5184.
8. Y.-S. Fu, J. Li, J. Li, *Nanomaterials* 9 (2019) 359.
9. S. Sood, S. Kumar, A. Umar, A. Kaur, S. K. Mehta, S. K. Kansal, *J. Alloys Compd.* 650 (2015) 193.
10. L. Gnanasekaran, R. Hemamalini, K. Ravichandran, *J. Saudi Chem. Soc.* 19 (2015) 589.
11. Q. Deng, W. Zhang, T. Lan, J. Xie, W. Xie, Z. Liu, Y. Huang, M. Wei, *Eur. J. Inorg. Chem.* 2018 (2018) 1506.
12. X. Xu, Z. Gao, Z. Cui, Y. Liang, Z. Li, S. Zhu, X. Yang, J. Ma, *ACS Appl. Mater. Interfaces* 8 (2016) 91.
13. X. Wang, S. Jiang, X. Huo, R. Xia, E. Muhire, M. Gao, *Nanotechnology* 29 (2018) 205702.
14. S. Wang, F. Wang, Z. Su, X. Wang, Y. Han, L. Zhang, J. Xiang, W. Du, N. Tang, *Catalysts* 9 (2019) 439.
15. J. Wang, W. Lin, H. Hu, C. Liu, Q Cai, S. Zhou, Y. Kong, *New J. Chem.* 45 (2021) 3067.
16. J.-H. Lee, S.-Y. Jeong, Y.-D. Son, S.-W. Lee, *Nanomaterials* 13 (2023) 1565.
17. X. Y. Hu, Y. T. Yu, D. D. Chen, W. C. Xu, J. Z. Fang, Z. Liu, R. Q. Li, L. Yao, J. J. Qin, Z. Q. Fang, *Chem. Eng. J.* 432 (2022) 15.
18. X. Zhu, Z. M. Dong, J. D. Xu, S. Y. Lin, J. Y. Liu, Z. P. Cheng, X. H. Cao, Y. Q. Wang, Y. H. Liu, Z. B. Zhang, *J. Alloys Compd.* 926 (2022) 10.
19. Z. W. Zhao, J. Y Fan, X. Y. Deng, J. Liu, *Chem. Eng. J.* 360 (2019) 1517.
20. S. Dutta, K. Das, K. Chakrabarti, D. Jana, S. K. De, S. De, *J. Phys. D Appl. Phys.* 49 (2016) 315107.
21. H. Zhao, Q. Fang, C. Chen, Z. Chao, Y. Tsang, Y. Wu, *Appl. Organomet. Chem.* 32 (2018) e4449.

22. M. Patel, J. S. Kim, B. S. Kim, J. Kim, *Mater. Lett.* 218 (2018) 123.
23. J. Xu, H. Guo, Y. Hao, J. Tian, Y. Liu, H. Ren, W. Yang, *J. Mater. Chem. C* 11 (2023) 15624.
24. L. Su, S. Lei, L. Liu, L. Liu, Y. Zhang, S. Shi, X. Yan, *J. Mater. Chem. A* 6 (2018) 9997.
25. H. Liu, W. Shen, X. Chen, *RSC Adv.* 9 (2019) 24519.
26. S. Dhara, D. Jariwala, S. Das, *Nanoscopy and Nanospectroscopy*, CRC Press, Boca Raton, FL (2023) 97.
27. G. Song, H. Sun, J. Chen, Z. Chen, B. Liu, Z. Liu, S. Cong, Z. Zhao, *Anal. Chem.* 94 (2022) 5048.
28. H. Yu, Z. Song, Q. Liu, X. Ji, J. Liu, S. Xu, H. Kan, B. Zhang, J. Liu, J. Jiang, L. Miao, H. Liu, *Sens. Actuators B Chem.* 248 (2017) 1029.
29. M. Epifani, E. Comini, R. Díaz, T. Andreu, A. Genç, J. Arbiol, P. Siciliano, G. Faglia, J. R. Morante, *ACS Appl. Mater. Interfaces* 6 (2014) 16808.
30. B. Zhang, X. He, C. Yu, G. Liu, D. Ma, C. Cui, Q. Yan, Y. Zhang, G. Zhang, J. Ma, Y. Xin, *Chin. Chem. Lett.* 33 (2022) 1337.
31. S. Ni, Z. Fu, L. Li, M. Ma, Y. Liu, *Colloids Surf. A Physicochem. Eng. Asp.* 649 (2022) 129475.
32. F. Guo, H. Sun, X. Huang, W. Shi, C. Yan, *J. Chem. Technol. Biotechnol.* 95 (2020) 2684.
33. P. Gündo˘gmu ̧s, J. Park, A. Öztürk, *Ceram. Int.* 46 (2020) 21431.
34. Y. Li, Q. Zhang, Y. Lu, Z. Song, C. Wang, D. Li, X. Tang, X. Zhou, *Ceram. Int.* 48 (2022) 1306.
35. Q. Yang, J. An, Z. Xu, S. Liang, H. Wang, *Colloids Surf. A Physicochem. Eng. Asp.* 614 (2021) 126161.
36. L. Liu, J. Huang, H. Yu, J. Wan, L. Liu, K. Yi, W. Zhang, C. Zhang, *Chemosphere* 282 (2021) 131049.
37. S. Xue, C. Wu, S. Pu, Y. Hou, T. Tong, G. Yang, Z. Qin, Z. Wang, J. Bao, *Environ. Pollut.* 250 (2019) 338.
38. A. Ray, A. Roy, S. Saha, S. Das, *Intechopen*, London (2018) https://doi.org/10.5772/intechopen.80298.
39. C. D. Lokhande, D. P. Dubal, O.-S. Joo, *Current Appl. Phys.* 11 (2011) 255.
40. J. Noh, C. M. Yoon, Y. K. Kim, J. Jang, *Carbon* 116 (2017) 470.
41. P. R. Reshma, P. Anees, S. Sinha, A. Dasgupta, K. Ganesan, A. K. Prasad, S. Dhara, *J. Phys. Chem. Lett.* 12 (2021) 9821.
42. S. Pawar, *Progress and Prospects in Nanoscience Today*. Novan Science Publishers, Inc., New York, USA (2020) ISBN: 978-1-53617-293-5.
43. S. A. Delbari, L. S. Ghadimi, R. Hadi, S. Farhoudian, M. Nedaei, A. Babapoor, A. S. Namini, Q. Van Le, M. Shokouhimehr, M. S. Asl, M. M. Mohammadi, *J. Alloys Compd.* 857 (2021) 15828.
44. F. Zheng, C. Xi, J. Xu, Y. Yu, W. Yang, P. Hu, Y. Li, Q. Zhen, S. Bashir, J. L. Liu, *J. Alloys Compd.* 772 (2019) 933.
45. J. Zhao, Y. Tian, A. Liu, L. Song, Z. Zhao, *Mater. Sci. Semicond. Processing* 96 (2019) 78.
46. J. H. Zheng, R. M. Zhang, X. G. Wang, P. F. Yu, *Res. Chem. Intermediates* 44 (2018) 5569.
47. B. Varshney, M. J. Siddiqui, A. H. Anwer, M. Z. Khan, F. Ahmed, A. Aljaafari, H. H. Hammud, A. Azam, *Sci. Rep.* 10 (2020) 11032.
48. H. T. Tan, X. Rui, H. Yu, W. Liu, C. Xu, Z. Xu, C. Xu, H. H. Hng, *ACS Nano* 8 (2014) 4004.

49. D. Yu, Z. Zhang, Y. Meng, Y. Teng, Y. Wu, X. Zhang, Q. Sun, W. Tong, X. Zhao, X. Liu, *Inorg. Chem. Front.* 5 (2018) 597.
50. R. Basu, S. Ghosh, S. Bera, A. Das, S. Dhara, *Sci. Rep.* 9 (2019) 4621.
51. R. Basu, S. Dhara, *J. Nanotech. Nanomat.* 1 (2020) 92.
52. W. Cai, R. Kumar Kankala, M. Xiao, N. Zhang, X. Zhang, *Appl. Surf. Sci.* 528 (2020) 146921.
53. G. Huang, C. Li, X. Sun, J. Bai, *New J. Chem.* 41 (2017) 8977.
54. J. Yus, Y. Bravo, A. J. Sanchez-Herencia, B. Ferrari, Z. Gonzalez, *Electrochimica Act* 308 (2019) 363.
55. O. C. Pore, A. V. Fulari, V. G. Parale, H. H. Park, R. V. Shejwal, V. J. Fulari, G. M. Lohar, *J. Porous Mater.* 29 (2022) 1991.
56. X. Liang, X. Zhang, W. Liu, D. Tang, B. Zhang, G. Ji, *J. Mater. Chem. C* 4 (2016) 6816.
57. K. Wang, J. Yang, J. Zhu, L. Li, Y. Liu, C. Zhang, T. Liu, *J. Mater. Chem. A* 5 (2017) 11236.
58. K. Chanda, S. Thakur, S. Maiti, A. Acharya, T. Paul, N. Besra, S. Sarkar, A. Das, K. Sardar, K. K. Chattopadhyay, *AIP Conf. Proc.* 1953 (2018) 030138.
59. M. A. A. M. Abdah, N. H. N. Azman, S. Kulandaivalu, Y. Sulaiman, *Mater. Design.* 186 (2019) 108199.
60. S. T. Navale, Z. B. Yang, C. Liu, P. J. Cao, V. B. Patil, N. S. Ramgir, R. S. Mane, F. J. Stadler, *Sens. Actuators B Chem.* 255 (2018) 1701.
61. T. Sugahara, L. Alipour, Y. Hirose, E. Yusufu, J.-I. Nakamura, H. Ono, N. Harada, K. Suganuma, *ACS Appl. Elec. Mater.* 2 (2020) 1670.
62. N. Mintcheva, P. Srinivasan, J. B. B. Rayappan, A. A. Kuchmizhak, S. Gurbatov, S. A. Kulinich, *Appl. Surf. Sci.* 507 (2019) 145169.
63. M. Epifani, S. Kaciulis, A. Mezzi, T. Zhang, J. Arbiol, P. Siciliano, A. Landström, I. Concina, A. Moumen, E. Comini, C. Xiangfeng, *Sens. Actuators B Chem.* 319 (2020) 128338.
64. N. Chen, D. Deng, Y. Li, X. Liu, X. Xing, X. Xiao, Y. Wang, *Sci. Rep.* 7 (2017) 7692.
65. K. Li, Y. Luo, L. Gao, T. Li, G. Duan, *ACS Appl. Mater.* 12 (2020) 16792.
66. N. Chen, D. Deng, Y. Li, X. Xing, X. Liu, X. Xiao, Y. Wang, *RSC Adv.* 6 (2016) 49692.
67. X. Wang, Y. Li, Z. Li, S. Zhang, X. Deng, G. Zhao, *Sens. Actuators B Chem.* 301 (2019) 127019.
68. P. Chaudhari, S. Mishra, *Measurement* 90 (2016) 468.
69. Z. Wang, M. Hu, Y. Wang, X. Liu, Y. Qin, *J. Alloys Compd.* 665 (2016) 173.
70. W. Wang, F. Li, N. Zhang, C. Liu, J. Zhou, D. Liu, S. Ruan, *Sens. Actuators B Chem.* 351 (2022) 130931.
71. D.-S. Lee, S.-D. Han, S.-M. Lee, J.-S. Huh, D.-D. Lee, *Sens. Actuators B Chem.* 65 (2000) 331.
72. Z. Han, J. Ren, J. Zhou, S. Zhang, Z. Zhang, L. Yang, C. Yin, *Int. J. Hyd. Energy* 45 (2020) 7223.
73. M. Abbaspoor, M. Aliannezhadi, F. S. Tehrani, *Opt. Mater.* 121 (2021) 111552.
74. Y. Zhang, C. Wu, B. Xiao, L. Yang, A. Jiao, K. Li, T. Chen, R. Zhan, Z. Huang, H. Lin, *Sens. Actuators B Chem.* 369 (2022) 132247.
75. Y. Li, Z. Hua, Y. Wu, Y. Zeng, Z. Qiu, X. Tian, M. Wang, E. Li, *Sens. Actuators B Chem.* 265 (2018) 249.
76. Z. Zhang, M. Haq, Z. Wen, Z. Ye, L. Zhu, *Appl. Surf. Sci.* 434 (2018) 891.
77. J.-Y. Shen, L. Zhang, J. Ren, J.-C. Wang, H.-C. Yao, Z.-J. Li, *Sens. Actuators B Chem.* 239 (2017) 597.

78. A. T. Güntner, N. A. Sievi, S. J. Theodore, T. Gulich, M. Kohler, S. E. Pratsinis, *Anal. Chem.* 89 (2017) 10578.
79. J. Sun, L. Sun, N. Han, J. Pan, W. Liu, S. Bai, Y. Feng, R. Luo, D. Li, A. Chen, *Sens. Actuators B Chem.* 285 (2019) 68.
80. X. Jie, D. Zeng, J. Zhang, K. Xu, J. Wu, B. Zhu, C. Xie, *Sens. Actuators B Chem.* 220 (2015) 201.
81. P.-G. Su, S.-L. Peng, *Talanta* 132 (2015) 398.
82. S. B. Kulkarni, Y. H. Navale, S. T. Navale, F. J. Stadler, N. S. Ramgir, V. B. Patil, *Sens. Actuators B Chem.* 288 (2019) 279.
83. J. Sun, X. Shu, Y. Tian, Z. Tong, S. Bai, R. Luo, D. Li, A. Chen, *Sens. Actuators B Chem.* 238 (2017) 510.

5 Heterostructures with Transition Metal Oxides

5.1 INTRODUCTION

Bulk transition metal oxides (TMOs) show outstanding physical properties, namely, magnetism, superconductivity, and multi-ferocity, which are of potential scientific interest and are adept for providing elucidation to pioneering electronics, energy, security, and environmental issues. At the same time, atomic-scale superlattices of TMOs have demonstrated various new physical concepts of condensed matter physics.[1] In TMOs, the transition metal ions possess an unfilled *d*-shell electronic configuration showing spin, orbital, and charge degrees of freedom. They are also correlated, leading to many-body phenomena.[2]

Very often, heterostructures of two distinct materials lead to functional properties different from the individual constituents mediated by interfacial interactions. The majority of these heterostructures are formed using structurally similar materials, namely, compound semiconductors, perovskite oxides, and van der Waals-bonded materials, among others.[3-6] However, heterostructures comprising different structural materials and having distinct interfaces also show similar behaviour and promote the possibility of including large varieties of material combinations that can be multi-dimensional.

In this chapter, we discuss the properties and applications of heterostructures comprising layers of either two-dimensional (2D) layered materials of transition metal dichalcogenides (TMDs),[7-10] graphene,[7,11-13] and metal carbide, nitrides, and carbonitrides (known as MXenes)[14,15] on TMOs. Reported modulation doping of 2D-TMD/TMO heterostructure demonstrated charge carrier injection useful for carrier transport in the defect-free 2D layer, giving rise to high-quality nanoelectronic devices.[8,16] The 2D graphene/TMO heterostructures are reported for high photoreponsivity, advanced energy storage, and solar cell applications.[17] Analogous to graphene, 2D black phosphorus on the $SrTiO_3$ surface showed enhanced photoresponsivity, leading to the application of the heterojunction as a programmable photoconductive switch.[18] MXene/TMO heterostructures found applications in capacitors, rechargeable batteries,[14] and capacitive deionization (CDI) process used for desalination based on electrosorption methodology.[19]

DOI: 10.1201/9781003623694-5

5.2 TMD/TMO HETEROSTRUCTURES

Atomically thin monolayer (ML) TMDs are direct bandgap semiconductors with high electron mobility and excitonic binding energy exhibiting excellent optoelectronic properties at room temperature. TMD-based devices are mostly used on SiO_2 substrates.[20-24] However, the van der Waals interaction between ML-TMDs and functional TMOs ($LaAlO_3$ or $SrTiO_3$) is demonstrated to modify the properties of the former. Photoluminescence (PL) studies are used to verify the van der Waals interaction by measuring the variation of PL intensity from ML-MoS_2 because of the charge transfer from the TMO substrate to MoS_2.[25] In another report, the altered optical contrast in the visible range corresponding to the ferroelectric domain of the periodically polarized lithium niobite (PPLN) film substrate consequential to the charge transfer between MoS_2 and PPLN provided direct evidence of the MoS_2-TMO surface interaction.[26]

MoS_2, likewise graphene, is considered as an electrode material and is controllable by the electric field when interfaces with a ferroelectric TMO, leading to the formation of a heterostructure ferroelectric tunnel junction device, namely, MoS_2/$BaTiO_3$/$SrRuO_3$ by utilizing the polarization of the ferroelectric TMO.[27] An asymmetric switching behaviour of the MoS_2/$BaTiO_3$/$SrRuO_3$ tunnel junction showed (Figure 5.1a-e) modulation of the electrical conduction at MoS_2/$BaTiO_3$ interface.[28] A heterostructure-based MoS_2/ferroelectric TMO device was demonstrated in the MoS_2 channel transistor with its gate connected to an external Pb(Zr,Ti)O_3 capacitor.

In this MoS_2/PZT transistor device, strong absorption of light in the ferroelectric layer induces a change in polarization-dependent switching as a cooperative behaviour through optical modulation (Figure 5.1f,g).[29] The PZT, as a ferroelectric layer among TMOs and Al-doped hafnium oxide among high-dielectric-constant materials were tested because of their compatibility with conventional Si-processing. In an innovative technique, modulation doping is demonstrated in 2D semiconductor/TMO heterostructures with different charge neutrality levels (CNLs). Carrier can be either injected into or extracted from the 2D semiconductor layer without perturbing the structural defects introduced by dopant atoms. Layer thickness-dependent negative trions are studied in MoS_2 by both electron injection into the active layer using TiO_2 film with a high CNL and extracting electrons using MoO_3 film with a low CNL.[8] The modulation doping process using TMO is compatible with Si-based semiconducting fabrication processes.

The molybdenum dichalcogenide (MoX_2) layered structure is sensitive to many gas atmospheres near room temperature, being advantageous with respect to TMOs operating generally at higher temperatures.[9] However, similar to TMOs, the MoX_2 layers are not selective in different chemical environments. $MoTe_2$ is reported to be an extremely sensitive environmental gas or chemical species. Material degradation and lack of systematic studies for $MoTe_2$- and $MoSe_2$-based sensors make the reported results in the literature unreliable.[30,31] On the other hand, the operating temperature of dispersed SnO_2 nanoparticles on the MoS_2 layer for 50 ppm ethanol sensing was reduced to 280 C with a response (R_{Gas}/R_{Air}) of 50.[32] The reduction of operating

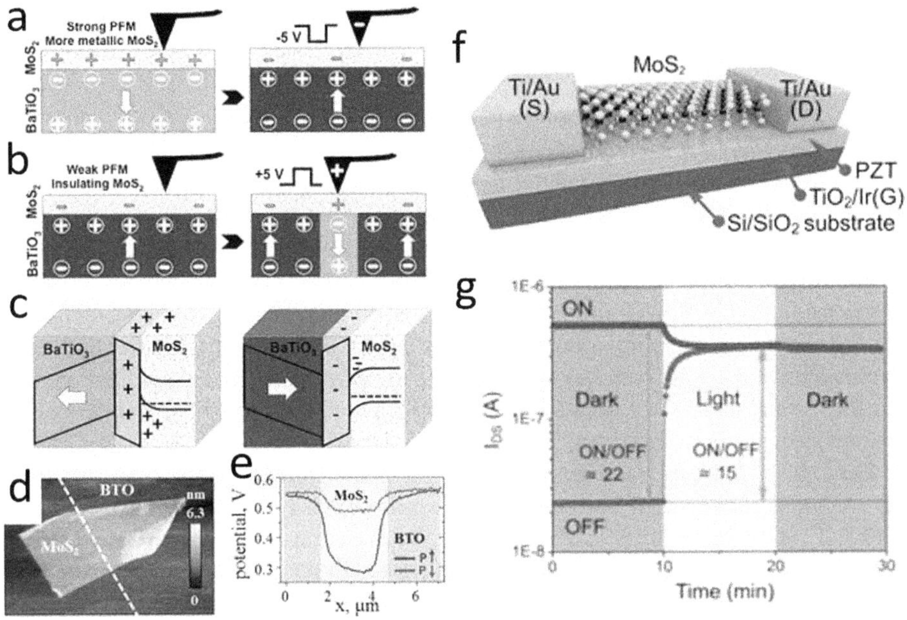

FIGURE 5.1 (**a-e**) Asymmetric switching behaviour of the MoS_2–$BaTiO_3$–$SrRuO_3$ tunnel junction: (**a**,**b**) schematics of complete switching from the downward to the upward state by a negative voltage pulse (**a**), and partial switching from the upward to the downward state by a positive voltage pulse (**b**); (**c**) simplified picture of polarization-induced band structure change; (**d**) topographic image of the MoS_2 flake on the surface of the 12 unit-cell-thick $BaTiO_3$ film; (**e**) surface potential profiles along the *dashed line* marked in (**d**) depending on the polarization directions. (**a-e**: Reproduced with permission from Reference [28]. Copyright 2017, American Chemical Society.) (**e**,**f**) MoS_2–$Pb(Zr,Ti)O_3$ ferroelectric FET: (**f**) device schematic and (**g**) effect of visible light illumination on the data retention characteristics. (**f**,**g**: Reproduced with permission from Reference [29]. Copyright 2015, American Chemical Society.)

temperature in the resistive sensing process was possible by prohibiting the agglomeration of SnO_2 nanoparticles. In another report, it is shown that the coating of SnO_2 nanocrystals stabilizes MoS_2 layers in the air.[33] The heterostructure comprising *n*-MoS_2 and *n*-SnO_2 showed a *p*-type behaviour in the FET configuration, enhancing the sensitivity, stability, and reproducibility of NO_2 with an optimum responsivity of 28% for 10 ppm of the pollutant in the air. Similar switching behaviour is also reported for MoS_2-decorated TiO_2 nanotube heterojunctions acting as resistive sensors with a response of 14.2 for 150 ethanol at 150°C.[34] However, stability study in these materials, including the most studied material, MoS_2, for recovery of adsorbed analytes is also not impressive. Thermal-assisted recovery prohibits the sensor operation at room temperature. Optically excited sensing mechanism using MoX_2/TMO heterostructures may be an alternative technique. However, significant material degradation under exposure to light may not be avoided.

5.3 GRAPHENE/2D TMO HETEROSTRUCTURES

5.3.1 Graphene/$SrTiO_3$ Thin-Film Heterostructures

In one of the studies, the transport behaviour of heterojunction graphene/epi-$SrTiO_3$ film was reported and was also compared with bulk $SrTiO_3$ in different orientations.[35] The transport characteristics for the heterojunction of graphene with thin film and bulk $SrTiO_3$, however, showed hysteretic conductance in the absence of characteristic quantum Hall behaviour. Characteristic Hall behaviour is only expected in the case of defect-free formation of heterojunction with the wrinkle-free graphene and stoichiometric formation of $SrTiO_3$ film.[36,37] The hysteretic behaviour is assumed to originate from either charge trapping or ferroelectric effects of the substrate. Similar hysteretic conductance behaviour was reported in another study.[38] One of the studies reported that the hysteresis was not because of the graphene intrinsic property, and the contribution might come from $SrTiO_3$.[35]

The role of oxygen vacancies was negated as the sample was annealed in an O environment. However, in both the reports, it was envisaged that the surface dipole of $SrTiO_3$, showing (Figure 5.2a,b) ferroelectric-like contribution, might be the origin of the observed hysteretic conductance behaviour. In addition, because of the high dielectric constant for $SrTiO_3$, the operating gate voltage was found to decrease compared with that of conventional Si-based devices. An interesting observation of saturation of the carrier density (Figure 5.2c,d) at an electric field of ~100 $kV \times cm^{-1}$ was made. In a separate study, the first report of Hall conductance was recorded in graphene/$SrTiO_3$ FET with a significant reduction of the operating voltage of the device.[39] A very high quality epi-$SrTiO_3$ (300 nm) layer was formed on the atomically flat Nb-doped $SrTiO_3$ single-crystal substrate. In the magnetotransport measurement, graphene demonstrates the quantum Hall conductance till 200 K (Figure 5.2e,f).[39] The high dielectric constant and the small thickness of the $SrTiO_3$ epilayer are the keys to achieving quantum conductance of graphene, even for the miniaturized device. Ambiguity-free dielectric properties of the layer adjacent to graphene, showing quantized conductance and being a quantity of the charge, can be measured by avoiding contributions from dead layer and electrodes used in direct capacitance measurement.[40] Moreover, in the graphene/$SrTiO_3$ heterojunction, graphene acts as an active electrode, and the electric field along the thin TMO film can result in the redox process of introducing electron-devouring O vacancies. The small thickness of the TMO thin film can accommodate a large electric field suitable for the development of the redox-process-induced hysteretic conductance of the graphene, leading to neuromorphic device applications.[41-43]

In a heterojunction of graphene/$SrTiO_3$ (90 nm) thin film, hysteretic quantum Hall conductance was first reported by Kang *et al.*[44] Linear conductance is dominant in the low field regime, and hysteresis develops systematically with the electric field exceeding the threshold of 110 $kV \times cm^{-1}$, inducing redox activity in the graphene layer. The O vacancies start to accumulate at the interface. The field-induced trapping sites accumulate carriers, preventing injection to the graphene layer and leading to saturation and hysteresis in the conductance. The opposite happens on the allocation of the negative field, which initiates the annihilation of O vacancies, restoring the

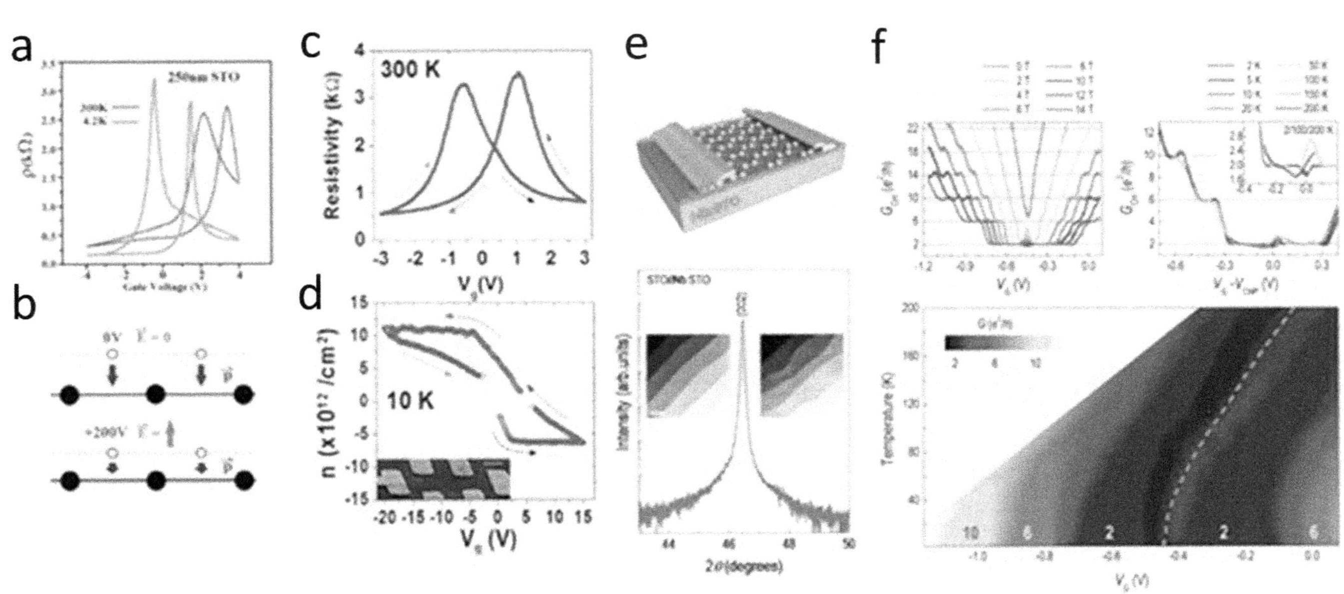

FIGURE 5.2 (**a**,**b**) Hysteretic conductance behaviour of graphene on $SrTiO_3$ thin film (250 nm). The scenario of the surface dipole of the $SrTiO_3$ thin film has been employed to understand the ferroelectric-like substrate phenomena. (**a**,**b**: Reproduced from Reference [35]. Creative Commons. Copyright 2014, Nature Publishing Group.) (**c**,**d**) Hysteretic conductance behaviour of graphene on $SrTiO_3$ thin film (300 nm) with a notable reduction of operating voltage. Although the hysteresis can be understood in terms of charge trapping, a curious saturation of the carrier density is observed at $\approx 100\ kV \times cm^{-1}$. (**c**,**d**: Reproduced from Reference [38]. Creative Commons. Copyright 2014, Nature Publishing Group.) (**e**) Schematic of monolayer graphene on $SrTiO_3$ thin-film device. To exclude extrinsic effects, the crystalline and atomically flat substrate is prepared. (**f**) Channel magnetotransport quantization as a function of gate voltage at 2 K under various magnetic fields and under 14 T magnetic field at various temperatures. The temperature-dependent shift of the charge-neutrality point (grey line) arises which is attributed to the temperature-dependent change of the dielectric nature of the substrate underneath. (Reproduced with permission from Reference [39]. Copyright 2016, American Chemical Society.)

charge neutrality. Quantum Hall conductance is observed with field-sensitive hysteresis application of a magnetic field. As discussed earlier, a quantitative relationship between the dielectric properties and the O vacancy layer thickness offers a methodology to investigate the intrinsic properties in the TMO layer.

A reconfigurable multifunctional device is studied in graphene/$LaAlO_3$/$SrTiO_3$ heterostructure consisting of large bandgap TMOs ($LaAlO_3$ and $SrTiO_3$) due to its fascinating metallic interface.[45-47] Polarization-induced metal-to-insulator transition (MIT) was observed with respect to the thickness of $LaAlO_3$ thin film.[46,48] Close to the critical thickness of $LaAlO_3$ (4 u.c.), MIT was locally switched using conductive atomic force microscope (c-AFM) lithography.[49] A graphene/$LaAlO_3$/$SrTiO_3$-based FET was reported by using c-AFM lithography.[50] The similarly configured high-mobility heterostructure FET was studied for its magnetoconductance characteristics.[51] The anomalous Hall effect and weak anti-localization effects showed quantum Hall properties for a large temperature range up to room temperature.

5.3.2 Graphene/Ferroelectric Heterostructures

For the planar device configuration, graphene/$Pb(Zr_xTi_{1-x})O_3$ (PZT) heterostructure devices were reported using graphene on top of PZT films with a back electrical contact. Electrical properties can be studied on the surface of the PZT substrate by forming source and drain contacts to the top graphene layer. Ferroelectricity in the PZT layer can be modulated in the FET configuration.[52-58] Graphene can also be used as a sensor for monitoring the ferroelectric state of PZT, with most of the changes occurring in the conductance of the ferroelectric switching region.[59-61] An anomalous hysteresis direction leading to anti-hysteresis showing a variation of the conductance of graphene when the gate bias is applied in a range more than coercive voltages corresponding to saturated polarization states was reported in the study of graphene/PZT heterostructure FET device.[53] A more complicated anti-hysteresis was reported in graphene/PZT transistors with a possible role of ferroelectric polarization switching of a capacitor in PZT, leading to the carrier modulation of graphene.[56] The modulation of the conductance of graphene is appraised as the memory device in several reports.

For other TMO-based ferroelectrics, the hysteresis direction of graphene on $PbTiO_3$/$SrTiO_3$ superlattices varied depending on the temperature and the gate bias.[62] The role of the surface state of the substrate with interfacial charge traps in this regard was understood as adsorbates from the environment and an intrinsic defect on the surface of the substrate. Graphene/$Ba_{1-x}Sr_xTiO_3$ thin film heterostructure showed modulation of the quantum Hall effect of graphene by invoking the ferroelectric polarization with a crossover between typical hysteresis and anti-hyteresis as the temperature varied.[63]

Unlike the thin-film PZT substrate, graphene on $[Pb(Mg_{1/3}Nb_{2/3})O_3]_{1-x}$-$[PbTiO_3]_x$ (PMN-PT) single-crystal showed typical hysteresis direction for the CVD-synthesized graphene.[64] However, the graphene/hexagonal-BN/PMN-PT device showed a transition between typical and anti-hysteresis depending on the voltage-sweep ranges.[65] As discussed earlier, interfacial charge traps play a major role in manifesting the anti-hysteresis behaviour in graphene/PZT and graphene/PMN-PT systems. Sluggish variation (extensive relaxation time) of the conductance of graphene as a function of the time after switching off the polarization is considered indirect evidence for the role of charge trap at interfaces.[61,66]

For the vertical device configuration, graphene/$BiFeO_3$/Pt heterostructure is also reported with conducting graphene as a substitution of the top indium-tin-oxide (ITO) contact layer for the application of $BiFeO_3$-based ferroelectric photovoltaic cell.[67] In another report, heterostructure Au/$BiFeO_3$/graphene/SiO_2/Si (substrate) was reported as vertical tunnelling device providing gate-bias tunable tunnelling current.[68]

Graphene/$BaTiO_3$/(La,Sr)MnO_3 tunnelling device structure is reported as a next-generation potential nonvolatile two-terminal memory device with a large on/off ratio depending on the polarization direction of $BiTiO_3$ and can clearly show the device operation on the environmental conditions.[69] Piezoelectric force microscopy (PFM) was used for flexoelectric domain switching in graphene/$BaTiO_3$/(La,Sr)MnO_3 heterostructure.[70] The graphene/ferroelectric TMO heterostructure can also be investigated in terms of the structural property variation revealed by Raman spectroscopy. The strain of graphene on graphene/ferroelectric TMO heterostructure can be modulated by the piezoelectricity of the PMN-PT substrate and monitored using a Raman signal.[71,72] In another configuration, graphene/PPLN can be used for the gate-dependent peak shift of the Raman G-band where different polarizations from opposite ferroelectric domains induce different amounts of charge from the substrate to graphene layer as gate control.[73]

5.3.3 Graphene/Single Elemental TMO Heterostructures

Novel optical and electronic device platform was reported in 2D graphene/TiO_2 heterostructures grown by depositing C60 on transition metal Ti substrate adopting a combined molecular beam epitaxy and O_2 intercalation methods.[11] The heterostructure showed 25 times higher photosensitivity than that reported for graphene in the visible light spectrum. Coulomb interactions at the interface promote optical transitions between O_{2p} orbitals of the valence band of TiO_2 and C_{2p} orbitals of the conduction band of graphene, showing an improved responsivity in the heterostructure. Additionally, 2D graphene/TiO_2 heterostructure was demonstrated as a bottom-gated graphene FET device with graphene as a channel and TiO_2 as a gate dielectric layer. Photocatalytic activity of titania nanoparticle/epilayer graphene on Ir(111) was reported to enhance with the intercalation of a 2D layer of $TiO_{1.5}$ between graphene and Ir substrate.[74] The observed enhancement was attributed to the dissimilar doping level induced in graphene by the $TiO_{1.5}$ layer and by the shifts of the Ti *d*-band of the titania nanoparticles.

A 2D graphene–TiO_2 sandwich structure grown in the reduction–hydrolysis technique showed the photocatalytic activity of CO_2 to useful hydrocarbons of CH_4 and C_2H_6 in water vapour, indicating important applications of graphene for selectively catalytic C–C coupling reaction.[75] Another sandwich-structured TiO_2–Pt–graphene hybrid showed enhanced electrochemical performance because of synergistic and resilient metal-support interaction.[76] (3-Aminopropyl)trithoxysilane was used to functionalize TiO_2 nanoparticles, which were subsequently wrapped by graphene, having potential for photocatalytic application.[77] Graphene-coated TiO_2 nanospheres are reported as photocatalysts for efficiently decomposing rhodamine B with an efficiency of up to 91% in 90 min, which is significantly higher than 65% reported for TiO_2 nanospheres alone.[78]

The incorporation of graphene increases the conductivity of VO_2 electrodes, which can be operated as high-performance electrochemical capacitors.[79] The hybrid structure shows enhanced optical response[80] and demonstrates robust electrodes for lithium-ion batteries (LIBs) and sodium-ion batteries (SIBs).[81] The use of an N-doped graphene-VO_2 nanosheet resulted in a three-dimensional (3D) flower morphology with a high surface area and improved electrical conductivity with the graphene, leading to better electrochemical performance.[82] In another study, the incorporation of graphene sheets into V_2O_5 nanoribbons increased electrical conductivity and conserved the water molecules between the two layers of V_2O_5. The intercalated layer helps the Li^+ diffusion significantly improve electrochemical performance.[83] V_2O_5 quantum dots (QDs)/graphene is also reported as an important cathode material for use in long-life rechargeable LIBs.[84,85] VO_2-graphene hybrid materials are studied for the fabrication of thermochromic films for reducing room temperature by cutting electromagnetic spectra above the IR range. VO_2 crystals were grown on graphene, and the film was relocated to a flexible substrate, demonstrating a novel flexible thermochromic film.[86]

Hybridized graphene-Cr_2O_3, with its achieved stability, was reported to demonstrate that oxygen reduction reaction (ORR) played an important role in renewable energy technologies involving water splitting and fuel cells.[87] It was also found to be applicable to energy storage.[88]

For other TMOs, the hybrid Mn_3O_4–graphene was reported for CO_2 adsorption,[89] ORR,[90] for catalytic oxidation and adsorption of elementary mercury,[91] for enhancing electrochemical performance for supercapacitors,[92-94] and photocatalytic activity of aqueous organics[95] and methylene blue.[96]

Ultrathin 2D MnO_2-graphene hybrids and graphene-wrapped MnO_2 are reported as superior-quality flexible planar supercapacitors.[97,98] MnO with low conversion potential, low-voltage hysteresis (<0.8 V), and high density (5.43 $g \cdot cm^{-3}$) find challenges in LIB applications.

The nontoxicity of Fe and the economic availability of Fe_2O_3 make it a component for energy storage. Fe_2O_3 is an attractive anode material for LIBs with a high theoretical capacity value of 1007 $mAh \cdot g^{-1}$,[99] which is three times larger than that of graphite. Electrode pulverization of Fe_2O_3 and rapid capacity decay during the cycling process, originating because of large specific volume change and particle aggregation, are removed by creating hybrids of Fe_2O_3-graphene having superior performance with respect to flexibility and electrical conductivity.[100] Graphene hydrogel supporting Fe_2O_3 was used as an advanced supercapacitor electrode material.[101] Composites of graphene and Fe_2O_3 found important applications in LIBs.[102-105] Core-shell hollow nanomaterial of γ-Fe_2O_3@graphene hybrid, prepared through the Kirkendall process, exhibited superior performance as an anode material for LIBs.[106] Because of adsorbent properties, monolithic Fe_2O_3 graphene hybrids were reported for removing As.[107] In another study, graphene/γ-Fe_2O_3 hybrid aerogels were studied for biocatalytic transformation.[108] A novel hybrid structure of Fe@Fe_2O_3 core-shell nanoparticle–graphene showed reversible Li storage.[109]

The 3D morphology of N-doped graphene-aerogel-supported Fe_3O_4 nanoparticle is reported as efficient cathode catalyst for ORR in alkaline electrodes,[110] which is even more significant than N-doped carbon black or N-doped graphene for its high effective surface area, in addition to exhibiting a higher current density and higher electron transfer, lower formation of H_2O_2, and better durability. The formation of porous iron oxide ribbons on graphene surfaces was found to be an application in superior-performing lithium storage.[111] Hybrid paper of graphene–Fe_3O_4 nanoparticle showed an actuation strain 56% higher than pristine graphene paper.[112] In a similar study, free-standing hybrid papers of graphene–Fe_3O_4 demonstrated magnetically actuated switching.[113]

Similar to other hybrid TMOs, Co_3O_4–graphene hybrids showing high oxygen reduction activity have also been studied as an anode for LIBs.[114,115] Co_3O_4–N-doped graphene showed better stability to Pt in alkaline solution than that of Co_3O_4–graphene hybrids with similar catalytic activity to the former.[116] Three-dimensional Co_3O_4/flocculent graphene hybrids on Ni foam were reported for supercapacitor applications, and the morphology activated synergistic improvement in electrochemical performance.[117] Symmetric supercapacitors of Co_3O_4 nanocluster on vertically aligned graphene were demonstrated for nonenzymatic glucose detection using the hybrid electrodes on a micropipette tip.[118] N-doped graphene/Co_3O_4 exhibited selective oxidation of organic alcohols and olefins[119] LIBs,[120] ORR,[121] act as catalysts for water electrolysis.[122] The Co_3O_4–graphene hybrid showed catalytic properties for heterogeneous activation of peroxymonosulfate in decomposing phenol.[123] Blunt-edge Co_3O_4 nanocube with low surface energy crystal plane on N-doped graphene showed higher catalytic activity toward oxygen evolution reaction (OER) in an alkaline medium. Graphene-encapsulated Co_3O_4 nanoclusters showed high reversible capacity of 1000 $mAh \cdot g^{-1}$ over 130 cycles and is superior to pristine Co_3O_4.[124] Naocrystalline CoO-graphene demonstrated a high-performance anode material with long cycle life, high reversible specific capacity, and excellent rate capability for LIB applications by avoiding pulverization and aggregation upon cycling.[125] Co/CoO–graphene hybrid has similar catalytic activity of nanocluster Pt with better stability than carbon-supported Pt nanoclusters for the ORR in an alkaline solution.[126] Porous-graphene-wrapped CoO nanoparticles showed higher OER activity in the porous structure of the hybrid, providing a high electroactive surface area.[127] Enzymeless glucose detection was achieved using a CoO–graphene hybrid sensor.[128] The same nonenzymatic sensor is used for the determination of carbofuran and carbaryl in vegetables and fruits.[129]

NiO is an intrinsically *p*-type wide bandgap semiconductor that is used as a catalyst, antiferromagnetic material, electrochemical capacitor, and battery cathode. Similar to other TMO-graphene hybrid systems, NiO–graphene hybrids are used for LIB applications.[130-132] Monolayer graphene/NiO nanosheet composite and 3D NiO/ultrathin-derived graphene are also reported for supercapacitor application.[133-135] In another application, the NiO–graphene hybrids reported reducing gas (H_2, NH_3, and

NO_2) sensing.[136] A nonenzymatic glucose sensor was also reported using graphene nanosheet–NiO hybrids in combination with DNA.[137] Graphene-wrapped NiO hybrids were utilized for supercapacitors,[138] and an electrochemical pseudocapacitor material for prospective energy storage applications.[139,140]

Both cuprous and cupric oxides are popular as *p*-type semiconductors and are useful for forming *p-n* junctions with a majority of oxide or nitride semiconductors, which are intrinsically *n*-type. The *p*-type Cu_2O semiconductor is envisaged for application in solar energy conversion, sensing, and photocatalytic degradation. Graphene-wrapped Cu_2O nanocubes demonstrated improved electro-oxidation of glucose and electrochemical stability.[141] The porous 3D Cu_2O nanowire-graphene hybrids demonstrated superior reducing NO_2 gas sensing. CuO-graphene composites,[141,142] and graphene-wrapped CuO hybrids[143] proved to be suitable as anode material for LIBs. A bipolar resistive switching device using CuO and multilayer graphene hybrid was reported utilizing the interlayer electronic interaction at the hybrid interface. The reversibility of bipolar resistive switching is functionalized by the O_2 intake capacity of the multilayer graphene.[144] CuO-graphene nanostructures were reported for nonenzymatic glucose[145] and humidity[146] sensing. The same hybrid structure was used for CO_2 mineralization,[147] as supercapacitors,[148] and electrodes in pseudocapacitor materials.[149] TMOs like Mn_3O_4, Fe_3O_4, and Co_3O_4, however, are significantly less conductive materials, so their use as hybrids with graphene poses a challenge in LIB applications.

In an exclusive study, the heterostructure made of epitaxial VO_2 on the bilayer of graphene on Al_2O_3 (0001) substrates showed a sharp temperature window for MIT.[13] With an enhanced value of on/off ratio, the heterostructure demonstrated improved transport properties as oxygen stoichiometry is better maintained with the prevention of O vacancy formation in the graphene interlayer. In a recent study, graphene layers were either mechanically or wet transferred on top of thin VO_2 layers grown on an Al_2O_3 substrate to avoid strain in the VO_2 layer originating from lattice mismatch and to study the effect of VO_2 on the graphene layer.[150] A nominal increase in the MIT temperature in graphene/VO_2 was observed as compared with that for pure VO_2. In the absence of a lattice-mismatch-induced strain effect, the increase in temperature of MIT may originate from a large difference in the thermal conductivity between graphene and VO_2.

5.4 MXENE/2D TMO HETEROSTRUCTURES

In the last two decades, a 2D layer of MXenes got reasonable attention because of tunable metal compositions, high electronic conductivity, low ion-diffusion barriers, and abundant active sites with functional termination groups intrinsic to the system.[151] Piling of the MXene layers along with preferential surface oxidation, however, restricts them to show their full potential in energy storage and water desalination applications.[152] Heterostructures of MXene/TMO have been reported to overcome these shortcomings. The hierarchical heterostructures can not only inhibit the piling of MXene sheets but also enhance the electrical properties and act as a barrier to the change in the volume of TMOs during the electrochemical activities. The MXene/TMO heterostructures act as novel electrode material for desalination and energy storage activities using the synergistic effect of active TMOs and conductive MXenes.

5.4.1 Li^+ Battery Applications

MXenes find potential applications as LIB anode materials. TiO_2 and SnO_2 nanowires grown on Ti_3C_2 MXene using van der Waals interactions in the self-assembly process were reported to demonstrate novel high-energy and power anode materials for application in LIBs.[14] The MXene layer assisted reversible electron and ion transport at the interface during the lithiation/de-lithiation process. In another report, MXene Ti_3C_2 showed a theoretical capacity of 447 $mA \times h \times g^{-1}$.[153] Fe_3O_4, a commonly used LIB anode, shows poor electrical conductivity and a substantial volume expansion.[152,154] However, a capacity of 782 $mA \times h \times g^{-1}$ was achieved by forming Fe_3O_4/MXene heterostructures. The carbonaceous (C)-Fe_3O_4/Ti_3C_2 heterostructure electrode exhibited an ultrahigh capacity of 1150 $mA \times h \times g^{-1}$ over 1000 cycles, where the Li^+ could diffuse efficiently through the amorphous carbon layer and few layered Ti_3C_2 nanosheets.[155] In a significant report, slow Li^+ kinetics and significant volume expansion upon cycling in Co_3O_4 were improved by using Co_3O_4 QDs/$Ti_3C_2T_x$ (T_x = surface functional groups) heterostructures, which exhibited a high-performance Li^+ storage capacity of 1005 $mA \times h \times g^{-1}$.[156] Nb_2O_5 with poor electronic conductivity was dealt with $Nb_2O_5/Nb_4C_3T_x$ where the active TMO nanoparticles could reduce the ion diffusion path, while the carbon-coupled 2D MXene enabled the electron transport. The heterostructure showed a capacity of 195 $mA \times h \times g^{-1}$.[157] Cost-effective and environment-friendly SnO_2 shows a significant change of volume leading to inefficient electrochemical cycling stability and Coulomb efficiency. To avoid the problem, the SnO_2@$Ti_3C_2T_x$ were reported to have a capacity of 810 $mA \times h \times g^{-1}$.[158] In addition, TiO_2, MnO_2, and Cu_2O have also formed heterostructure configurations of $TiO_2/Ti_3C_2T_x$, Cu_2O/Ti_2C, and MnO_2/MXene electrodes.[159-161]

5.4.2 Na^+ Battery Applications

Presently, SIBs have drawn attention because of the abundant low cost of Na^+ and better safety than while handling Li^+. However, the ion diffusion is slower for Na^+ than for Li^+ as the radius of the former is larger, along with the shortcomings of substantial electrode volume expansion for SIB during the electrochemical cycling process.[162] α-VO_x/VC heterostructures showed a capacity of 307 $mA \times h \times g^{-1}$.[163] In different reports, TiO_2/Ti_3C_2 heterostructures are reported for capacities of 153[164] and 101 $mA \times h \times g^{-1}$.[165]

5.4.3 Li-S Battery Applications

Li-S batteries are the one likely energy storage devices owing to the application of low-cost sulphur cathode and high energy density of 2600 $W \times h \times kg^{-1}$.[162,166,167] However, the use of Li-S batteries is limited by low electronic conductivity, meagre

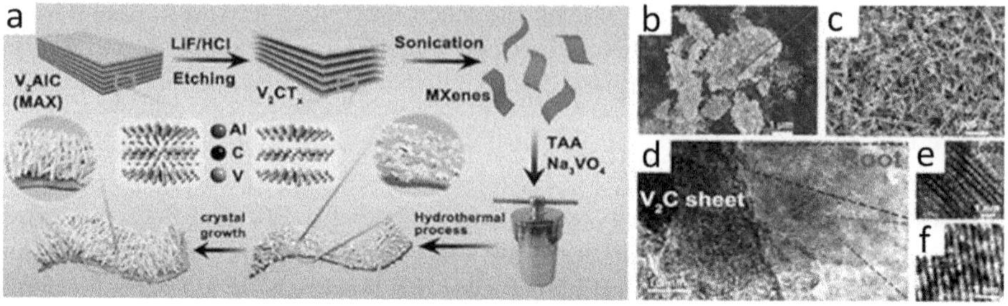

FIGURE 5.3 (**a**) Schematic diagram of VO_2/V_2CT_x composite, (**b, c**) medium- and high-magnification of the VO_2/V_2CT_x composite, and (**d-f**) the HRTEM image of VO_2/V_2CT_x composite. (Reproduced with permission from Reference [172]. Copyright 2019, American Chemical Society.)

electrochemical cyclability, and dissolution of poly-sulphide.[168,169] Ti_4O_7, V_2O_3, and MnO_2 TMOs with polar chemical bonds can inhibit poly-sulphide transporting.[170,171] However, the low conductivity of TMO limits the chemical reaction between S and TMOs. TiO_2@MXene hybrids exhibited a capacity of 680 $mA \times h \times g^{-1}$.[167] $VO_2/V_2C/S$ heterostructures reported (Figure 5.3) for a capacity of 855 $mA \times h \times g^{-1}$.[172] MXene@$TiO_2$/S hybrids showed a capacity value of 612 $mA \times h \times g^{-1}$,[173] and MnO_x/MXene heterostructures reported a capacity of 650 $mA \times h \times g^{-1}$.[172]

5.4.4 Supercapacitor Applications

Supercapacitors have drawn attention as representative electrochemical energy storage devices owing to their high power density, superior rate capability, and significantly long lifecycle.[175] In application as supercapacitor, WO_3-Ti_3C_2 hybrid structures exhibited a capacity of 566 $mA \times h \times g^{-1}$.[176] At the same time, MnO_2/MXene heterostructures showed a capacity value of 448 $mA \times h \times g^{-1}$.[177] Heterostructures of ε-$MnO_2/Ti_3C_2T_x$ reported for capacity of 130 $mA \times h \times g^{-1}$.[178] MnO_2-Ti_3C_2 hybrids measured for a capacity of 130 $mA \times h \times g^{-1}$. The $MoO_3/Ti_3C_2T_x$ heterostructure exhibited a capacity value of 151 $mA \times h \times g^{-1}$.[179] Hybrid structure of NiO-$Ti_3C_2T_x$ demonstrated a capacity of 85 $mA \times h \times g^{-1}$.[180]

5.4.5 Capacitive Deionization

A new seawater desalination process using the CDI technique has drawn significant consideration as being cheap, simple to operate, and compatible with environmental conditions.[19,181,182] Carbon, being cheap, abundant, and with reasonably high electrical conductivity, is a popular CDI electrode. However, the low desalination capacity of C restricts its future use. Heterostructures of $TiO_2/Ti_3C_2T_x$ showed a desalination performance of 23.8 $mg \times g^{-1}$.[183] In another report, $W_{18}O_{49}/Ti_3C_2$ heterostructures exhibited a desalination capability of 29.25 $mg \times g^{-1}$.[184] Moreover, a very high value of desalination performance 44 $mg \times g^{-1}$ was reported in Fe_3O_4@Ti_3C_2 hybrid structures.[185]

REFERENCES

1. J. Chakhalian, J. W. Freeland, A. J. Millis, C. Panagopoulos, J. M. Rondinelli, *Rev. Mod. Phys.* 86 (2014) 1189.
2. M. Imada, A. Fujimori, Y. Tokura, *Rev. Mod. Phys.* 70 (1998) 1039.
3. H. Kroemer, Quasi-Electric Fields and Band Offsets: Teaching Electrons New Tricks in *Nobel Lectures* (*ed.* G. Ekspong), World Scientific Publishing Co., Singapore (2000).
4. *Nat. Mater.* 11 (2012) 91. https://doi.org/10.1038/nmat3244
5. P. Zubko, S. Gariglio, M. Gabay, P. Ghosez, J.-M. Triscone, *Annu. Rev. Condens. Matter Phys.* 2 (2011) 141.
6. K. S. Novoselov, A. Mishchenko, A. Carvalho, A. H. Castro Neto, *Science* 353 (2016) 9439.
7. K. T. Kang, J. Park, D. Suh, W. S. Choi, *Adv. Mater.* 31 (2019) 1803732.
8. K. Xu, Y. Wang, Y. Zhao, Y. Chai, *J. Mater. Chem. C* 5 (2017) 376.
9. D. Zappa, *Materials* 10 (2017) 1418.
10. D. Zappa, V. Galstyan, N. Kaur, H. M. M. M. Arachchige, O. Sisman, E. Comini, *Anal. Chim. Acta* 1039 (2018) 1.
11. J. Park, T. Back, W. C. Mitchel, S. S. Kim, S. Elhamri, J. Boeckl, S. B. Fairchild, R. Naik, A. A. Voevodin, *Sci. Rep.* 5 (2015) 3332.
12. A. Jana, E. Scheer, S. Polarz, *Beilstein J. Naotechnol.* 8 (2017) 688.
13. H. Cao, X. Yan, Y. Li, L. Stan, W. Chen, N. P. Guisinger, H. Zhou, D. D. Fong, *Appl. Phys. Lett.* 121 (2022) 081601.
14. Y.-T. Liu, P. Zhang, N. Sun, B. Anasori, Q.-Z. Zhu, H. Liu, Y. Gogotsi, B. Xu, *Adv. Mater.* 30 (2018) 1707334.
15. W. Xi, J. Jin, Y. Zhang, R. Wang, Y. Gong, B. He, H. Wang, *Nanoscale* 14 (2022) 11923.
16. D. Wang, X.-B. Li, H.-B. Sun, *Nano Lett.* 21 (2021) 6298.
17. J. Azadmanjiri, V. K. Srivastava, P. Kumar, J. Wang, A. Yu, *J. Mater. Chem. A* 6 (2018) 13509.
18. A. L. Yeats, Y. Pan, A. Richardella, P. J. Mintun, N. Samarth, D. D. Awschalom, *Sci. Adv.* 1 (2015) e1500640.
19. S. Porada, R. Zhao, A. van der Wal, V. Presser, P. M. Biesheuvel, *Prog. Mater. Sci.* 58 (2013) 1388.
20. C. W. Rischau, X. Lin, C. P. Grams, D. Finck, S. Harms, J. Engelmayer, T. Lorenz, Y. Gallais, B. Fauqué, J. Hemberger, K. Behnia, *Nat. Phys.* 13 (2017) 643.
21. B. Radisavljevic, A. Radenovic, J. Brivio, V. Giacometti, A. Kis, *Nat. Nanotechnol.* 6 (2011) 147.
22. Z. Y. Yin, H. Li, H. Li, L. Jiang, Y. M. Shi, Y. H. Sun, G. Lu, Q. Zhang, X. D. Chen, H. Zhang, *ACS Nano* 6 (2012) 74.
23. N. R. Pradhan, D. Rhodes, S. M. Feng, Y. Xin, S. Memaran, B. H. Moon, H. Terrones, M. Terrones, L. Balicas, *ACS Nano* 8 (2014) 5911.
24. D. Ovchinnikov, A. Allain, Y. S. Huang, D. Dumcenco, A. Kis, *ACS Nano* 8 (2014) 8174.
25. S. Larentis, B. Fallahazad, E. Tutuc, *Appl. Phys. Lett.* 101 (2012) 223104.
26. Y. Li, Z. Qi, M. Liu, Y. Wang, X. Cheng, G. Zhang, L. Sheng, *Nanoscale* 6 (2014) 15248.
27. F. Liu, C. Zhu, L. You, S.-J. Liang, S. Zheng, J. Zhou, Q. Fu, Y. He, Q. Zeng, H. J. Fan, L. K. Ang, J. Wang, Z. Liu, *Adv. Mater.* 28 (2016) 7768.
28. T. Li, P. Sharma, A. Lipatov, H. Lee, J. W. Lee, M. Y. Zhuravlev, T. R. Paudel, Y. A. Genenko, C. B. Eom, E. Y. Tsymbal, A. Sinitskii, A. Gruverman, *Nano Lett.* 17 (2017) 922.
29. A. Lipatov, P. Sharma, A. Gruverman, A. Sinitskii, *ACS Nano* 9 (2015) 8089.

30. H. C. Diaz, R. Chaghi, Y. Ma, M. Batzill, *2D Mater.* 2 (2015) 044010.
31. G. Mirabelli, C. McGeough, M. Schmidt, E. K. McCarthy, S. Monaghan, I. M. Povey, M. McCarthy, F. Gity, R. Nagle, G. Hughes, *J. Appl. Phys.* 120 (2016) 125102.
32. H. Yan, P. Song, S. Zhang, Z. Yang, Q. Wang, *RSC Adv.* 5 (2015) 79593.
33. S. Cui, Z. Wen, X. Huang, J. Chang, J. Chen, *Small* 11 (2015) 2305.
34. P. X. Zhao, Y. Tang, J. Mao, Y. X. Chen, H. Song, J. W. Wang, Y. Song, Y. Q. Liang, X. M. Zhang, *J. Alloys Compd.* 674 (2016) 252.
35. R. Sachs, Z. Lin, J. Shi, *Sci. Rep.* 4 (2015) 3657.
36. J. D. Baran, C. Eames, K. Takahashi, M. Molinari, M. S. Islam, S. C. Parker, *Chem. Mater.* 29 (2017) 7364.
37. S. A. Lee, H. Jeong, S. Woo, J.-Y. Hwang, S.-Y. Choi, S.-D. Kim, M. Choi, S. Roh, H. Yu, J. Hwang, S. W. Kim, W. S. Choi, *Sci. Rep.* 6 (2016) 23649.
38. S. Saha, O. Kahya, M. Jaiswal, A. Srivastava, A. Annadi, J. Balakrishnan, A. Pachoud, C.-T. Toh, B.-H. Hong, J.-H. Ahn, T. Venkatesan, B. Özyilmaz, *Sci. Rep.* 4 (2015) 6173.
39. J. Park, H. Kang, K. T. Kang, Y. Yun, Y. H. Lee, W. S. Choi, D. Suh, *Nano Lett.* 16 (2016) 1754.
40. H.-C. Li, W. Si, A. D. West, X. X. Xi, *Appl. Phys. Lett.* 73 (1998) 464.
41. R. Waser, M. Aono, *Nat. Mater.* 6 (2007) 833.
42. A. Sawa, *Mater. Today* 11 (2008) 28.
43. R. Waser, R. Dittmann, G. Staikov, K. Szot, *Adv. Mater.* 21 (2009) 2632.
44. K. T. Kang, H. Kang, J. Park, D. Suh, W. S. Choi, *Adv. Mater.* 29 (2017) 1700071.
45. A. Ohtomo, H. Y. Hwang, *Nature* 427 (2004) 423.
46. N. Nakagawa, H. Y. Hwang, D. A. Muller, *Nat. Mater.* 5 (2006) 204.
47. D. Pesquera, M. Scigaj, P. Gargiani, A. Barla, J. Herrero-Martín, E. Pellegrin, S. M. Valvidares, J. Gázquez, M. Varela, N. Dix, J. Fontcuberta, F. Sánchez, G. Herranz, *Phys. Rev. Lett.* 14 (2014) 156802.
48. S. Thiel, G. Hammerl, A. Schmehl, C. W. Schneider, J. Mannhart, *Science* 313 (2006) 1942.
49. C. Cen, S. Thiel, G. Hammerl, C. W. Schneider, K. E. Andersen, C. S. Hellberg, J. Mannhart, J. Levy, *Nat. Mater.* 7 (2008) 298.
50. M. Huang, G. Jnawali, J.-F. Hsu, S. Dhingra, H. Lee, S. Ryu, F. Bi, F. Ghahari, J. Ravichandran, L. Chen, P. Kim, C.-B. Eom, B. D'Urso, P. Irvin, J. Levy, *APL Mater.* 3 (2015) 062502.
51. G. Jnawali, M. Huang, J.-F. Hsu, H. Lee, J.-W. Lee, P. Irvin, C.-B. Eom, B. D'Urso, J. Levy, *Adv. Mater.* 29 (2017) 1603488.
52. X. D. Wang, Y. Chen, G. J. Wu, J. L. Wang, B. B. Tian, S. Sun, H. Shen, T. Lin, W. D. Hu, T. T. Kang, M. H. Tang, Y. G. Xiao, J. L. Sun, X. J. Meng, J. H. Chu, *Nanotechnology* 29 (2018) 134002.
53. A. Rajapitamahuni, J. Hoffman, C. H. Ahn, X. Hong, *Nano Lett.* 13 (2013) 4374.
54. H. Lu, A. Lipatov, S. Ryu, D. J. Kim, H. Lee, M. Y. Zhuravlev, C. B. Eom, E. Y. Tsymbal, A. Sinitskii, A. Gruverman, *Nat. Commun.* 5 (2014) 5518.
55. N. Park, H. Kang, J. Park, Y. Lee, Y. Yun, J. H. Lee, S. G. Lee, Y. H. Lee, D. Suh, *ACS Nano* 9 (2015) 10729.
56. X. Hong, A. Posadas, K. Zou, C. H. Ahn, J. Zhu, *Phys. Rev. Lett.* 102 (2009) 136808.
57. X. Hong, J. Hoffman, A. Posadas, K. Zou, C. H. Ahn, J. Zhu, *Appl. Phys. Lett.* 97 (2010) 033114.
58. E. B. Song, B. Lian, S. M. Kim, S. Lee, T.-K. Chung, M. Wang, C. Zeng, G. Xu, K. Wong, Y. Zhou, H. I. Rasool, D. H. Seo, H.-J. Chung, J. Heo, S. Seo, K. L. Wang, *Appl. Phys. Lett.* 99 (2011) 042109.

59. Z. Yi, N. Guang-Xin, B. Sukang, C. Chun-Xiao, K. Orhan, T. Chee-Tat, K. Hye Ri, I. Danho, Y. Ting, A. Jong Hyun, H. Byung, Hee, Ö. Barbaros, *EPL* 93 (2011) 17002.
60. C. Baeumer, S. P. Rogers, R. Xu, L. W. Martin, M. Shim, *Nano Lett.* 13 (2013) 1693.
61. A. Lipatov, A. Fursina, T. H. Vo, P. Sharma, A. Gruverman, A. Sinitskii, *Adv. Electron. Mater.* 3 (2017) 1700020.
62. S. Lee, Y. Lee, *Carbon* 126 (2018) 176.
63. C. Ma, R. Lu, G. Hu, J. Han, M. Liu, J. Li, J. Wu, *ACS Appl. Mater. Interfaces* 9 (2017) 4244.
64. X. W. Zhang, D. Xie, J. L. Xu, C. Zhang, Y. L. Sun, Y. F. Zhao, X. Li, X. M. Li, H. W. Zhu, H. M. Chen, T. C. Chang, *Carbon* 93 (2015) 384.
65. S. P. Rogers, R. J. Xu, S. Pandya, L. W. Martin, M. Shim, *J. Phys. Chem. C* 121 (2017) 7542.
66. M. H. Yusuf, B. Nielsen, M. Dawber, X. Du, *Nano Lett.* 14 (2014) 5437.
67. W. Jie, Y. Y. Hui, N. Y. Chan, Y. Zhang, S. P. Lau, J. Hao, *J. Phys. Chem. C* 117 (2013) 13747.
68. W. J. Jie, J. H. Hao, *Nanoscale* 10 (2018) 328.
69. Y. Zang, D. Xie, X. Wu, Y. Chen, Y. Lin, M. Li, H. Tian, X. Li, Z. Li, H. Zhu, T. Ren, D. Plant, *Appl. Phys. Lett.* 99 (2011) 132904.
70. S. G. Yuan, Z. B. Yang, C. Xie, F. Yan, J. Y. Dai, S. P. Lau, H. L. W. Chan, J. H. Hao, *Adv. Mater.* 28 (2016) 10048.
71. H. Lu, B. Wang, T. Li, A. Lipatov, H. Lee, A. Rajapitamahuni, R. Xu, X. Hong, S. Farokhipoor, L. W. Martin, *Nano Lett.* 16 (2016) 6460.
72. F. Ding, H. Ji, Y. Chen, A. Herklotz, K. Dörr, Y. Mei, A. Rastelli, O. G. Schmidt, *Nano Lett.* 10 (2010) 3453.
73. W. Jie, Y. Y. Hui, Y. Zhang, S. P. Lau, J. Hao, *Appl. Phys. Lett.* 102 (2013) 223112.
74. D. D. Angelis, F. Presel, N. Jabeen, L. Bignardi, D. Lizzit, P. Lacovig, S. Lizzit, T. Montini, P. Fornasiero, D. Alfe, A. Baraldi, *Carbon* 157 (2020) 350.
75. W. Tu, Y. Zhou, Q. Liu, S. Yan, S. Bao, X. Wang, M. Xiao, Z. Zou, *Adv. Funct. Mater.* 23 (2013) 1743.
76. B. Y. Xia, B. Wang, H. B. Wu, Z. Liu, X. Wang, X. W. Lou, *J. Mater. Chem.* 22 (2012) 16499.
77. J. S. Lee, K. H. You, C. B. Park, *Adv. Mater.* 24 (2012) 1084.
78. J. Zhang, Z. Zhu, Y. Tang, X. Feng, *J. Mater. Chem. A* 1 (2013), 3752.
79. L. Deng, G. Zhang, L. Kang, Z. Lei, C. Liu, Z.-H. Liu, *Electrochim. Acta* 112 (2013) 448.
80. H. Kim, Y. Kim, T. Kim, A-R. Jang, H. Y. Jeong, S. H. Han, D. H. Yoon, H. S. Shin, D. J. Bae, K. S. Kim, W. S. Yang, *Nanoscale* 5 (2013) 2632.
81. D. Chao, C. Zhu, X. Xia, J. Liu, X. Zhang, J. Wang, P. Liang, J. Lin, H. Zhang, Z.-X. Shen, H.-J. Fan, *Nano Lett.* 15 (2015) 565.
82. C. Nethravathi, C. R. Rajamathi, M. Rajamathi, U. K. Gautam, X. Wang, D. Golberg, Y. Bando, *ACS Appl. Mater. Interfaces* 5 (2013) 2708.
83. Q. Liu, Z.-F. Li, Y. Liu, H. Zhang, Y. Ren, C.-J. Sun, W. Lu, Y. Zhou, L. Stanciu, E. A.nStach, J. Xie, *Nat. Commun.* 6 (2014) 6127.
84. C. Han, M. Yan, L. Mai, X. Tian, L. Xu, X. Xu, Q. An, Y. Zhao, X. Ma, J. Xie, *Nano Energy* 2 (2013) 916.
85. X. Rui, J. Zhu, D. Sim, C. Xu, Y. Zeng, H. Hng, T. M. Lim, Q. Y. Yan, *Nanoscale* 3 (2011) 4752.
86. H. Kim, Y. Kim, K. S. Kim, H. Y. Jeong, A-R. Jang, S. H. Han, D. H. Yoon, K. S. Suh, H. S. Shin, T. Kim, W. S. Yang, *ACS Nano* 7 (2013) 5769.

87. C. Wen, X. Gao, T. Huang, X. Wu, L. Xu, J. Yu, H. Zhang, Z. Zhang, J. Han, H. Ren, *Int. J. Hydrogen Energy* 11 (2016) 11099.
88. C. Song, Y. Gui, X. Xing, W. Zhang, *Mater. Chem. Phys.* 173 (2016) 460.
89. Z. Ding, L. Qing, C. QianYi, Z. YanChao, C. Yi, W. Tao, H. BaoHang, *Chin. Sci. Bull.* 57 (2012) 3059.
90. J. Duan, S. Chen, S. Dai, S. Z. Qiao, *Adv. Funct. Mater.* 24 (2014) 2072.
91. H. Xu, Z. Qu, C. Zong, W. Huang, F. Quan, N. Yan, *Environ. Sci. Technol.* 49 (2015) 6823.
92. L. Li, Z. Hu, Y. Yang, P. Liang, A. Lu, H. Xu, Y. Hu, H. Wu, *Chin. J. Chem.* 31 (2013) 1290.
93. Y. Hu, C. Guan, G. Feng, Q. Ke, X. Huang, J. Wang, *Adv. Funct. Mater.* 25 (2015) 7291.
94. X. Yang, Y. He, Y. Bai, J. Zhang, L. Kang, H. Xu, F. Shi, Z. Lei, Z.-H. Liu, *Electrochim. Acta* 188 (2016) 398.
95. Y. Yao, C. Xu, S. Yu, D. Zhang, S. Wang, *Ind. Eng. Chem. Res.* 52 (2013) 3637.
96. J. Qu, L. Shi, C. He, F. Gao, B. Li, Q. Zhou, H. Hu, G. Shao, X. Wang, J. Qiu, *Carbon* 66 (2014) 485.
97. L. Peng, X. Peng, B. Liu, C. Wu, Y. Xie, G. Yu, *Nano Lett.* 13 (2013) 2151.
98. J. Zhu, J. He, *ACS Appl. Mater. Interfaces* 4 (2012) 1770.
99. J. Chen, L. Xu, W. Li, X. Gou, *Adv. Mater.* 17 (2005) 582.
100. T. Wang, Z. Liu, M. Lu, B. Wen, Q. Ouyang, Y. Chen, C. Zhu, P. Gao, C. Li, M. Cao, L. Qi, *J. Appl. Phys.* 113 (2013) 024314.
101. Z. Ma, X. Huang, S. Dou, J. Wu, S. Wang, *J. Phys. Chem. C* 118 (2014) 17231.
102. M. Zhang, B. Qu, D. Lei, Y. Chen, X. Yu, L. Chen, Q. Li, Y. Wang, T. Wang, *J. Mater. Chem.* 22 (2012) 3868.
103. X.-Y. Xue, C.-H. Ma, C.-X. Cui, L.-L. Xing, *Solid State Sci.* 13 (2011) 1526.
104. X. Zhu, Y. Zhu, S. Murali, M. D. Stoller, R. S. Ruoff, *ACS Nano* 5 (2011) 3333.
105. L. Xiao, D. Wu, S. Han, Y. Huang, S. Li, M. He, F. Zhang, X. Feng, *ACS Appl. Mater. Interfaces* 5 (2013) 3764.
106. J. Hu, J. Zheng, L. Tian, Y. Duan, L. Lin, S. Cui, H. Peng, T. Liu, H. Guo, X. Wang, F. Pan, *Chem. Commun.* 51 (2015) 7855.
107. L. Li, G. Zhou, Z. Weng, X.-Y. Shan, F. Li, H.-M. Cheng, *Carbon* 67 (2014) 500.
108. L. Chen, B. Wei, X. Zhang, C. Li, *Small* 9 (2013) 2331.
109. C. Wu, H. Zhang, Y.-X. Wu, Q.-C. Zhuang, L.-L. Tian, X.-X. Zhang, *Electrochim. Acta* 134 (2014) 18.
110. Z.-S. Wu, S. Yang, Y. Sun, K. Parvez, X. Feng, K. Müllen, *J. Am. Chem. Soc.* 134 (2012) 9082.
111. S. Yang, Y. Sun, L. Chen, Y. Hernandez, X. Feng, K. Müllen, *Sci. Rep.* 2 (2012) 427.
112. J. Liang, Y. Huang, J. Oh, M. Kozlov, D. Sui, S. Fang, R. H. Baughman, Y. Ma, Y. Chen, *Adv. Funct. Mater.* 21 (2011) 3778.
113. J. Liang, Y. Xu, D. Sui, L. Zhang, Y. Huang, Y. Ma, F. Li, Y. Chen, *J. Phys. Chem. C* 114 (2010) 17465.
114. H. Kim, D.-H. Seo, S.-W. Kim, J. Kim, K. Kang, *Carbon* 49 (2011) 326.
115. J. Zhu, Y. Sharma, Z. Zeng, X. Zhang, M. Srinivasan, S. Mhaisalkar, H. Zhang, H. Hng, Q. Yan, *J. Phys. Chem. C* 115 (2011) 8400.
116. Y. Liang, Y. Li, H. Wang, J. Zhou, J. Wang, T. Regier, H. Dai, *Nat. Mater.* 10 (2011) 780.
117. C. Wu, Q. Shen, R. Mi, S. Deng, Y. Shu, H. Wang, J. Liu, H. J. Yan, *Mater. Chem. A* 2 (2014) 15987.
118. X. Wang, X. Dong, Y. Wen, C. Li, Q. Xiong, P. Chen, *Chem. Commun.* 48 (2012) 6490.

119. R. Nie, J. Shi, W. Du, W. Ning, Z. Hou, F.-S. Xiao, *J. Mater. Chem. A* 1 (2013) 9037.
120. M. Zhang, R. Li, X. Chang, C. Xue, X. Gou, *J. Power Sources* 290 (2015) 25.
121. T. Odedairo, X. Yan, J. Ma, Y. Jiao, X. Yao, A. Du, Z. Zhu, *ACS Appl. Mater. Interfaces* 17 (2015) 21373.
122. H. Jin, J. Wang, D. Su, Z. Wei, Z. Pang, Y. Wang, *J. Am. Chem. Soc.* 137 (2015) 2688.
123. Y. Yao, Z. Yang, H. Sun, S. Wang, *Ind. Eng. Chem. Res.* 51 (2012) 14958.
124. S. Yang, X. Feng, S. Ivanovici, K. Müllen, *Angew. Chem., Int. Ed.* 49 (2010) 8408.
125. Y. Sun, X. Hu, W. Luo, Y. Huang, *J. Phys. Chem. C* 116 (2012) 20794.
126. S. Guo, S. Zhang, L. Wu, S. Sun, *Angew. Chem. Int. Ed.* 51 (2012) 11770.
127. Y. Zhao, B. Sun, X. Huang, H. Liu, D. Su, K. Sun, G. Wang, *J. Mater. Chem. A* 3 (2015) 5402.
128. S. Ci, S. Mao, T. Huang, Z. Wen, D. A. Steeber, J. Chen, *Electroanalysis* 26 (2014) 1326.
129. M. Wang, J. Huang, M. Wang, D. Zhang, J. Chen, *Food Chem.* 151 (2014) 191.
130. X. H. Huang, P. Zhang, J. B. Wu, Y. Lin, R. Q. Guo, *Mater. Lett.* 153 (2015) 102.
131. Y. J. Mai, S. J. Shi, D. Zhang, Y. Lu, C. D. Gu, J. P. Tu, *J. Power Sources* 204 (2012) 155.
132. G. Zhou, D.-W. Wang, L.-C. Yin, N. Li, F. Li, H.-M. Cheng, *ACS Nano* 6 (2012) 3214.
133. B. Zhao, J. Song, P. Liu, W. Xu, T. Fang, Z. Jiao, H. Zhang, Y. Jiang, *J. Mater. Chem.* 21 (2011) 18792.
134. Y. Jiang, D. Chen, J. Song, Z. Jiao, Q. Ma, H. Zhang, L. Cheng, B. Zhao, Y. Chu, *Electrochim. Acta* 91 (2013) 173.
135. I. R. M. Kottegoda, N. H. Idris, L. Lu, J.-Z. Wang, H.-K. Liu, *Electrochim. Acta* 56 (2011) 5815.
136. L. T. Hoa, H. N. Tien, V. H. Luan, J. S. Chung, S. H. Hur, *Sens. Actuators B* 135 (2013) 701.
137. W. Lv, F.-M. Jin, Q. G. Guo, Q.-H. Yang, F. Kang, *Electrochim. Acta* 73 (2012) 129.
138. R. Kumar, R. K. Singh, R. Savu, P. K. Dubey, P. Kumar, S. A. Moshkalev, *RSC Adv.* 6 (2016) 26612.
139. H. Wang, H. S. Casalongue, Y. Liang, H. Dai, *J. Am. Chem. Soc.* 132 (2010) 7472.
140. H. Wang, J. T. Robinson, G. Diankov, H. Dai, *J. Am. Chem. Soc.* 132 (2010) 3270.
141. M. Liu, R. Liu, W. Chen, *Biosens. Bioelectron.* 45 (2013) 206.
142. Y. J. Mai, X. L. Wang, J. Y. Xiang, Y. Q. Qiao, D. Zhang, C. D. Gu, J. P. Tu, *Electrochim. Acta* 56 (2011) 2306.
143. W. Zhou, F. Zhang, S. Liu, J. Wang, X. Du, D. Yin, L. Wang, *RSC Adv.* 4 (2014) 51362.
144. B. Singh, B. R. Mehta, Govind, X. Feng, K. Müllen, *Appl. Phys. Lett.* 99 (2011) 222109.
145. C.-L. Sun, W.-L. Cheng, T.-K. Hsu, C.-W. Chang, J.-L. Chang, J.-M. Zen, *Electrochem. Commun.* 30 (2013) 91.
146. Z. Wang, Y. Xiao, X. Cui, P. Cheng, B. Wang, Y. Gao, X. Li, T. Yang, T. Zhang, G. Lu, *ACS Appl. Mater. Interfaces* 6 (2014) 3888.
147. J. W. Ko, S.-W. Kim, J. Hong, J. Ryu, K. Kang, C. B. Park, *Green Chem.* 14 (2012) 2391.
148. K. K. Purushothaman, B. Saravanakumar, I. M. Babu, B. Sethuraman, G. Muralidharan, *RSC Adv.* 4 (2014) 23485.
149. B. Zhao, P. Liu, H. Zhuang, Z. Jiao, T. Fang, W. Xu, B. Lu, Y. Jiang, *J. Mater. Chem. A* 1 (2013) 367.
150. K. Lerttraikul, W. Rattanasakuldilok, T. Pakornchote, T. Bovornratanaraks, I. Klanurak, T. Taychatanapat, L. Srathongsian, C. Seriwatanachai, P. Kanjanaboos, S. Chatraphorn, S. Kittiwatanakul, *Sci. Rep.* 14 (2024) 4545.

151. B. Anasori, M. R. Lukatskaya, Y. Gogotsi, *Nat. Rev. Mater.* 2 (2017) 16098.
152. Y. Wang, Y. Li, Z. Qiu, X. Wu, P. Zhou, T. Zhou, J. Zhao, Z. Miao, J. Zhou, S. Zhuo, *J. Mater. Chem. A* 6 (2018) 11189.
153. D. Er, J. Li, M. Naguib, Y. Gogotsi, V. B. Shenoy, *ACS Appl. Mater. Interfaces* 6 (2014) 11173.
154. Q. Li, J. Zhou, F. Li, Z. Sun, *Inorg. Chem. Front.* 7 (2020) 3491.
155. Q. Li, J. Zhou, F. Li, Z. Sun, *Appl. Surf. Sci.* 530 (2020) 147214.
156. Y. Zhao, C. Liu, R. Yi, Z. Li, Y. Chen, Y. Li, I. Mitrovic, S. Taylor, P. Chalker, L. Yang, C. Zhao, *Electrochim. Acta* 345 (2020) 136203.
157. C. J. Zhang, S. J. Kim, M. Ghidiu, M.-Q. Zhao, M. W. Barsoum, V. Nicolosi, Y. Gogotsi, *Adv. Funct. Mater.* 26 (2016) 4143.
158. J. Xiong, L. Pan, H. Wang, F. Du, Y. Chen, J. Yang, C. Zhang, *Electrochim. Acta* 268 (2018) 503.
159. Y. T. Liu, P. Zhang, N. Sun, B. Anasori, Q. Z. Zhu, H. Liu,Y. Gogotsi, B. Xu, *Adv. Mater.* 30 (2018) 1707334.
160. H. Zhang, H. Dong, X. Zhang, Y. Xu, J. Fransaer, *Electrochim. Acta* 202 (2016) 24.
161. J. Yan, L. Yu, D. Wang, W. Zhang, Z. Xiong, T. Nie, Z. Ji, X. Yan, *Adv. Mater.* 2 (2021) 4772.
162. X. Guo, X. Xie, S. Choi, Y. Zhao, H. Liu, C. Wang, S. Chang, G. Wang, *J. Mater. Chem. A* 5 (2017) 12445.
163. W. Zhang, J. Peng, W. Hua, Y. Liu, J. Wang, Y. Liang, W. Lai, Y. Jiang, Y. Huang, W. Zhang, H. Yang, Y. Yang, L. Li, Z. Liu, L. Wang, S. L. Chou, *Adv. Energy Mater.* 11 (2021) 2100757.
164. P. Wang, X. Lu, Y. Boyjoo, X. Wei, Y. Zhang, D. Guo, S. Sun, J. Liu, *J. Power Sources* 451 (2020) 227756.
165. C. Yang, Y. Liu, X. Sun, Y. Zhang, L. Hou, Q. Zhang, C. Yuan, *Electrochim. Acta* 271 (2018) 165.
166. H. Zhang, Q. Qi, P. Zhang, W. Zheng, J. Chen, A. Zhou, W. Tian, W. Zhang, Z. Sun, *ACS Appl. Energy Mater.* 2 (2018) 705.
167. X. T. Gao, Y. Xie, X. D. Zhu, K. N. Sun, X. M. Xie, Y. T. Liu, J. Y. Yu, B. Ding, *Small* 14 (2018) 1802443.
168. Y. Zhang, C. Ma, W. He, C. Zhang, L. Zhou, G. Wang, W. Wei, *Prog. Natl. Sci. Mater.* 31 (2021) 501.
169. Z. Xiao, Z. Li, X. Meng, R. Wang, *J. Mater. Chem. A* 7 (2019) 22730.
170. X. Liu, J. Q. Huang, Q. Zhang, L. Mai, *Adv. Mater.* 29 (2017) 1601759.
171. Y. Zhong, K. R. Yang, W. Liu, P. He, V. Batista, H. Wang, *J. Phys. Chem. C* 121 (2017) 14222.
172. Z. Wang, K. Yu, Y. Feng, R. Qi, J. Ren, Z. Zhu, *ACS Appl. Mater. Interfaces* 11 (2019) 44282.
173. S. Y. Qiu, C. Wang, Z. X. Jiang, L. S. Zhang, L. L. Gu, K. X. Wang, J. Gao, X. D. Zhu, G. Wu, *Nanoscale* 12 (2020) 16678.
174. C. Song, Q. Yan, T. Zhang, H. Lin, H. Ye, Q. Yao, S. Zhang, Y. Li, G. Wang, J. Yang Lee, *Chem. Eng. J.* 420 (2021) 130452.
175. W. Liu, K. Wang, C. Li, X. Zhang, X. Sun, J. Han, X.-L. Wu, F. Li, Y. Ma, *J. Mater. Chem. A* 6 (2018) 24979.
176. S. B. Ambade, R. B. Ambade, W. Eom, S. H. Noh, S. H. Kim, T. H. Han, *Adv. Mater. Interfaces* 5 (2018) 1801361.
177. Z. Bo, K. Yi, H. Yang, X. Guo, Z. Huang, Z. Zheng, J. Yan, K. Cen, K. Ostrikov, *J. Power Sources* 492 (2021) 229639.

178. R. B. Rakhi, B. Ahmed, D. Anjum, H. N. Alshareef, *ACS Appl. Mater. Interfaces* 8 (2016) 18806.
179. J. Zhu, X. Lu, L. Wang, *RSC Adv.* 6 (2016) 98506.
180. K. Zhang, G. Ying, L. Liu, F. Ma, L. Su, C. Zhang, D. Wu, X. Wang, Y. Zhou, *Materials* 12 (2019) 188.
181. Z.-H. Huang, Z. Yang, F. Kang, M. Inagaki, *J. Mater. Chem. A* 5 (2017) 470.
182. W. Xi, H. Li, *Environ. Sci. Nano* 7 (2020) 764.
183. H. Li, W. Xi, *J. Inorg. Mater.* 36 (2021) 283.
184. J. Liang, J. Yu, W. Xing, W. Tang, N. Tang, J. Guo, *Chem. Eng. J.* 435 (2022) 134922.
185. K. Wang, L. Chen, G. Zhu, X. Xu, L. Wan, T. Lu, L. Pan, *Desalination* 522 (2022) 115420.

6 Perovskite Oxides

Properties at 1D and 2D

6.1 INTRODUCTION

Perovskite oxides show a variety of functional properties, namely, superconductivity, piezoelectricity, pyroelectricity, switchable polarization, non-linear dielectric behaviour, magnetism, photocatalysis, as well as multiferroic properties essential for application in micro-electronic devices such as ultra-sensitive sensors, micro-actuators, infrared detectors, non-volatile memory, microwave phase filters, spintronic devices among others.[1-3] However, low-dimensional perovskite oxides, particularly 2D systems up to monolayer thickness, exhibit physical properties entirely different from their bulk properties. Recent research has emphasized that 2D materials like graphene and transition metal dichalcogenides (TMDs) exhibited exotic electronic and optical properties only when the layer thickness is reduced to a monolayer. Thus, studying the size-dependent physical properties of perovskite oxides is necessary to realize a new generation of electronic devices at reduced dimensions.

We discuss the growth and studies of novel physical properties of 1D perovskite oxide containing transition metals in the form of nanowires and nanotubes.[1,2,4] We also study the growth and exotic physical properties of transition metal containing 2D perovskite oxide down to a monolayer,[3,5,6] demonstrating physical properties entirely different from their bulk properties.

6.2 1D PEROVSKITE OXIDES

6.2.1 Growth Techniques

Template-assisted growth of perovskite oxide nanowires resulted in polycrystallinity in the samples. So various techniques of solgel template routes,[7] hydrothermal methods,[8-10] microwave-hydrothermal methods,[11] solvothermal method assisted by a polymer,[12,13] solution-phase decomposition of bimetallic alkoxide precursors,[14,15] molten salt technique[16,17] were adopted to grow the nanowires of $BaTiO_3$ (BTO), $PbTiO_3$ (PTO), $Pb(Zr,Ti)O_3$ (PZT), $SrTiO_3$ (STO), $LaFeO_3$ (LFO), $R_{1-x}A_xTO_3$ (R = rare earth, A = alkaline, T= transition metal) perovskites. Single-crystalline tetragonal PTO nanowires of a few tens of micrometres in length and varying diameters

DOI: 10.1201/9781003623694-6

of 10–200 nm were grown using the electrospinning method.[18] Apart from ferroelectric perovskite oxide nanowires, the mixed valence perovskite manganite nanowires find application because of their interesting properties of metal-to-insulator transition (MIT) coupled with ferromagnetic-paramagnetic (FM-PM) transition, charge and orbital ordering, and the coexistence of various electronic and magnetic phases originating from strong coupling among charge, spin, orbit, and lattice degree of freedom. Single crystalline $La_{1-x}Ca_xMnO_3$ (LCMO; $x = 0.2$) and $La_{1-x}Sr_xMnO_3$ (LSMO; $x = 0.175$) nanowires of diameter 30 and 60–70 nm, respectively, were synthesized using a simple solgel process combined with nanoporous alumina as a template.[19-21] Single crystalline orthorhombic LCMO ($x = 0.5$) nanowires of several tens of micrometres with diameter ~80 nm,[22] LSMO ($x = 0.5$) nanowires of 1.5-15 μm with diameter ~50-400 nm,[23] LBMO ($x = 0.5$) of several tens of micrometres with diameter ~30-150 nm,[24] and $Pr_{0.5}Ca_{0.5}MnO_3$ (PCMO)[25] nanowires of a few microns length and diameter around 50 nm were grown in the low-temperature hydrothermal process. Nanowires of $La_{1-x}Ba_xMnO_3$ (LBMO; $x = 0.5$)[26] with diameters 30–150 nm, LSMO ($x = 0.5$)[27] with diameters 50–400 nm, and LCMO ($x = 0.5$)[28] with diameter 80 nm and length of several tens of micrometres long were grown by hydrothermal process. Single crystalline PCMO ($x = 0.5$) with a diameter of ~50 nm and a length of a few microns were reported to grow in the same process.[25] Core-shell nanowires of LSMO ($x = 0.33$)@MgO were grown by pulsed-laser deposition (PLD) of an epitaxial shell of 10 nm LSMO on single crystal MgO nanowires.[29] In a similar process, LPCMO/MgO core-shell nanowires were synthesized with diameters of about tens of nanometres in two steps.[30]

Numerous perovskite oxide nanotubes of BTO,[31-35] STO,[35] $(Ba,Sr)TiO_3$,[36-38] PTO,[31-33,39-44] PZO,[45] PZT,[32,40,46-54] $SrBi_2Ta_2O_9$ (SBTO),[55,56] $Bi_{3.25}La_{0.75}Ti_3O_{12}$,[57] $BiFeO_3$ (BFO),[58-62] LCMO,[63-65] LSCO ($x = 0.4$)[66-68] were synthesized by using different chemical routes. Small isolated crystalline BTO and STO nanotubes of diameter 8–15 nm with wall thickness 4–8 nm were grown in low temperature hydrothermal process using TiO_2 nanotubes as a precursor.[69] Arrays of BTO and $(Ba_xSr_{1-x})TiO_3$ nanotubes were grown on Ti substrates in the hydrothermal process.[37,70] Polycrystalline PZT nanotubes with a wall thickness of ~10 nm showing AFM properties were fabricated within the porous alumina membrane (PAM) templates using PLD technique.[45] Array of PZT nanotubes were synthesized by a solgel electrodeposition process where PZT sol was taken by DC electrophoretic voltages into the PAM template channels.[46] PZT and Pt/PZT/Pt nanotubes showing ferroelectricity were grown by magnetron sputtering and PLD techniques.[49] Multiferroic BFO nanotube with an aspect ratio of 200 and diameter 240 nm (wall thickness ~40 nm) was synthesized using a template-assisted solgel technique.[58-62]

LCMO nanowires of a diameter in the range of 185–195 nm,[71] and in a separate report,[72] nanowire LCMO ($x = 0.2$) of diameter 30 nm were grown using anodized aluminium oxide (AAO) assisted solgel route. Nanowires of LSMO ($x = 0.33$), LCMO ($x = 0.33$), and $La_{(1-x/2)}Pr_xCa_{(1-x/2)}MnO_3$ ($x = 0.3$) were synthesized in the same route with a length of ~6-8 μm, and the average wall thickness of 45, 60, and 150 nm, respectively.[73] LSMO ($x = 0.4$) nanotubes and nanowires with diameters of 100 nm and 40–60 nm were synthesized in the AAO template-assisted solgel route.[74] LSMO

($x = 0.3$) nanowires were grown in the polymer template-assisted chemical solution route.[75]

Nanowire LCMO ($x = 0.33$) with an aspect ratio greater than 300 were fabricated using both optical and focused ion beam (FIB) lithography.[76] In a similar technique, $(La_{0.5}Pr_{0.5})_{0.67}$-$Ca_{0.33}MnO_3$ nanobridges of width 100 nm were also reported.[77] Quasi-1D arrays of LSMO ($x = 0.33$) nanoconstriction of 100 nm on STO(100) substrate were fabricated using nanosphere lithography with SiO_2 microspheres.[78] $La_{0.33}Pr_{0.34}Ca_{0.33}MnO_3$ (LPCMO) nanochanels were made from a single crystalline thin film using optical lithography.[79] Nanochanels (200–1000 nm) of LSMO ($x = 0.33$)[80] and 20 nm width nanobridges of LSMO ($x = 0.3$)[81] on STO(100) substrate were also reported using FIB lithography.[80] Nanobridges of LCMO ($x = 0.3$) with 500 nm width and 100 nm thickness were fabricated by electron beam lithography on STO(100) substrate.[82]

Nanofibers of BTO with an average diameter of <100 nm,[83] LSMO ($x = 0.33$) of diameters about 80–300 nm and 200 μm length,[84] and LMO with diameter in the range of 50–100 nm were grown in the electrospinning method.[85]

6.2.2 Physical Properties

The nanoscale ferroelectricity of a single 15 nm BTO nanowire with a coercive field for polarization reversal ~7 kV×cm^{-1} and 100 nm^2 of ferroelectric non-volatile polarization domain size (areal capacity of 1 terabit×cm^{-2}) was reported (Figure 6.1)

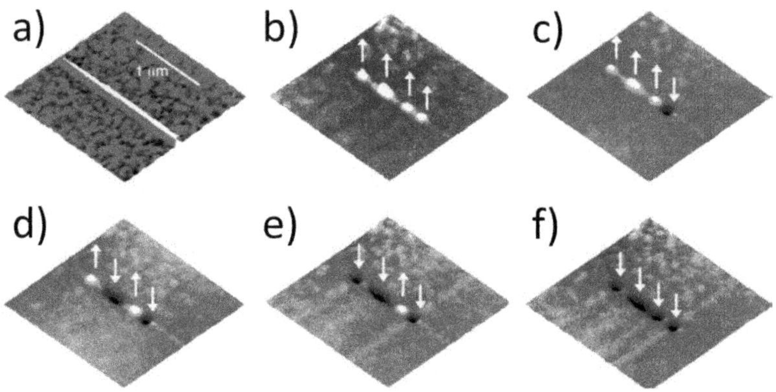

FIGURE 6.1 (**a**) 3D topographic image of a BTO nanowire with a diameter of 12 nm. (**b-f**) 3D EFM images showing that four distinct polarization domains can be independently manipulated by an external electric field. In these EFM images, the bright and dark colours correspond to a resonance frequency shift of +10 and −10 Hz, respectively, and *white arrows* indicate the polarization directions. The upward and downward polarization spots were written with $V_{tip} = -10$ V and $V_{tip} = +10$ V, respectively. The distance between the tip and the top surface of the nanowire was kept for 10 nm during the writing procedure and 35 nm during the reading procedure. (Reprinted with permission from Reference [87]. Copyright 2002, American Chemical Society.)

using electrostatic force microscopy (EFM).[86] The electric transport studies in LSMO ($x = 0.33$)@MgO core-shell nanowires showed a MIT at 325 K in the absence of a magnetic field.[29] The magnetoresistance (MR) decreased by 9.2% at 1 T as compared with zero field. It is found that in the PCMO nanowires, the charge ordering transition is suppressed, and a ferromagnetic phase is observed, whereas the antiferromagnetic transition disappears.[25] The ferro- and piezoelectric properties of Si template-grown PZT and BTO nanotubes using polymeric precursors,[32,33] were established by piezoelectric force microscopic (PFM) studies. PFM study of BFO nanotubes showed a significant decrease in d_{33} and the dielectric loss values up to 10^3 Hz and remained fairly constant till 10^6 Hz.[59] The values of d_{33} and dielectric loss at 10^5 Hz were 90 and 0.04, respectively, and comparable to literature values.[87] Superparamagnetism, originating from uncompensated surface spins prompted by the size effect, was observed in PAM template-assisted solgel grown BFO nanotube arrays.[60] The first report on the complete magnetization loop of a single LCMO ($x = 0.33$) was demonstrated using a Si micro-oscillator.[88] A FM-PM transition below 200 K was reported in $La_{(1-x/2)}Pr_xCa_{(1-x/2)}MnO_3$ ($x = 0.3$) nanotubes.[65]

A single crystalline LPCMO nanochannel, scaled down to the order of the inherent electronic phase separation, showed that MIT has originated from the first-order phase transition between ferromagnetic metal and charge-ordered insulator domains.[79] A 34% enhancement of MR value was reported in LCMO ($x = 0.33$) nanowire of diameter 150 nm because of the release of strain at the edges together with a destabilization of the insulating regions for probable use as functional spintronic devices.[76] An enhanced Curie temperature (T_C) was reported in orthorhombic LCMO ($x = 0.5$) nanowires because of the anisotropy induced by the contraction of the unit cell.[28] PCMO ($x = 0.5$) nanowires showed a reduction in the charge-ordered phase and disappearance of the antiferromagnetic phase.[25]

Kondo effect originating from blocked Mn spins at the grain boundaries was reported in $La_{0.67}Sr_{0.33}MnO_3$@MgO nanorod arrays with three-order higher resistance than that of thin films with enhanced scattering at the grain boundaries.[89] Higher coercivity (H_C) value was reported for the LPCMO nanowires than that of the bulk sample.[90] The higher H_C value observed in the nanowires was because of stronger domain wall pinning at the boundaries of the separated AFM and FM phases originated due to the electronic phase separation in the nanowires.[91]

6.2.3 Device Applications

PZT nanotubes showed promise as nanogenerators with the largest output voltage as high as 469 mV for having significantly higher piezoelectricity value of all the PZT materials (d_{33} = 71–289 pC×N^{-1}) such as ZnO (d_{33} = 12 pC×N^{-1}).[51] PZT nanotubes of wall thickness around 40 nm grown on *n*-type Si substrates exhibited a strong terahertz source with a broad peak in the novel frequency range of 2-8 THz in 0.2 ps because of an unusually large carrier concentration gradient in the nanotubes.[92] The frequency range is not covered in conventional semiconductor terahertz devices (ZnTe, or *p*-InAs) and was only reported for quantum wells.[93] The possible origin of high-frequency source is optical rectification within a surface accumulation layer in the optical pumping process. However, it may also be an electrically driven emission because of ferroelectric switching in the PZT. The observation is very interesting as it is considerably cheaper than the commonly used $Hg_{1-x}Cd_xTe$ films on CdZnTe substrates as a terahertz source.[94] The role of absorbed O_2 molecules onto the PZT nanotubes may also contribute to the terahertz source.[92]

An all-perovskite ferroelectric FET structure with CMR LCMO ($x = 0.2$) channel and gated by $Pb(Zr_{0.2}Ti_{0.8})O_3$ structure on Si (Figure 6.2) showed modulation of 20% after an electric field poling of 1.5×10^5 V×cm^{-1}.[95] A maximum modulation of 50% with 1 T field was observed near the MIT of the CMR channel, exhibiting the possibility of integrating the CMR-based ferroelectric FET on Si. An all-oxide MTJ device was made, for the first time, using LSMO ($x = 0.33$) as FM and STO as insulating layers.[96] In another report, a record tunnelling MR ratio of 1850% was also reported in a similar configuration.[97]

Nanotubes of PZT (aspect ratio 6.5) with top (inner) and bottom (outer) electrodes as Ru[98] and SBT (aspect ratio 250)[99] with Ru top and Pd bottom electrodes are reported as self-trenching conformal capacitors for non-volatile memory applications. In a metal–insulator–metal configuration, LSMO ($x = 0.33$) as an insulator layer was studied for the first time, using a scanning probe microscope conducting tip (metal) for the electronic properties of the high-resistance state and their dependence on temperature and magnetic field.[100] The study found the role of electronic effects as a major reason for charge depletion, even though mobile ions or ionic defects might

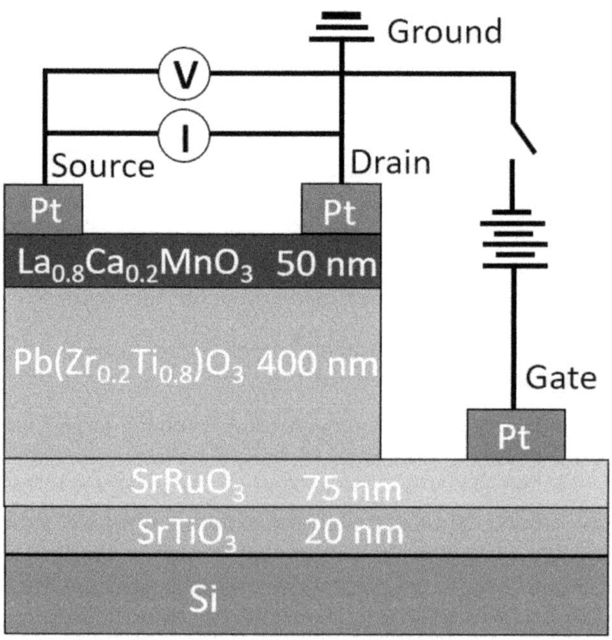

FIGURE 6.2 Schematic structure of the ferroelectric FET. (Reproduced with permission from Reference [96]. Copyright 2004, American Institute of Physics.)

affect the carrier transport, leading to the possible application of oxide-based non-volatile memory devices.

Free-standing ferroelectric nanotubes find other interesting applications, namely, externally activated piezoelectric syringes for drug delivery systems[101] and nano-fluidic devices.[102]

6.3 2D PEROVSKITE OXIDES

6.3.1 Growth Techniques

Monolayer growth of perovskite BFO and STO ultrathin films was reported by reactive molecular beam epitaxy (MBE) technique on a water-soluble $Sr_3Al_2O_6$ as the sacrificial buffer layer and transferring them to desired substrates.[3] The 2D layers of BTO, STO, PTO were reported by MBE[104,105] and PLD techniques.[105,106] Solution-based technique was also adopted to grow 2D $KNbO_3$ sheets.[107] PTO films of varying thicknesses were synthesized on $GdScO_3$(110) substrates using the PLD technique.[108] Thickness dependence of the ferroelastic domains of PTO films grown on $DyScO_3$(110) substrates was also studied.[109] Growth of PTO(001) films on MgO(001) and STO(001) substrates was analysed for the stability of domain structures having a fraction of domains in a poly-twin structure developed in the epitaxial heterostructures. Theoretical predictions are in good agreement with experimental results, with active PTO layers experiencing a tetragonal to orthorhombic or a cubic to tetragonal phase transformation. The maps for epitaxial $BaTiO_3$(001) films grown on various substrates were also included.[110] A toroidal state in STO/PTO/STO stack with the arrangement of alternating clockwise/counterclockwise polar vortices was reported.[5]

6.3.2 Physical Properties and Applications

In the monolayer growth process, BFO films showed unexpected and giant tetragonality and polarization when approaching the 2D limit (Figure 6.3).[3] The study also established that the crystalline order is independent of thickness for the free-standing ultrathin oxide films. Irrespective of either or both of the SrO- and TiO_2-terminal groups on the STO layer, an unusual quantum size effect was theoretically predicted in 2D STO with a lowering of bandgap than that reported for the bulk sample.[5] The origin of opposite behaviour was envisaged due to the formation of surface states of valence bands and the splitting of conduction bands. $KNbO_3$ nanosheets showed a strong anisotropy in ferroelectric polarization.[107] A sound ferroelectric state with the polarization along [100] induced by the surface effect and an improper hybrid ferroelectric state with the polarization along [110] was predicted in $SrVO_3$ and $CaFeO_3$ layers using density functional theory (DFT),[6] suggesting a new direction for fabricating 2D ferroelectric and multiferroic devices with superior properties.

The phase transition as a function of the number of unit cells in the confined PTO layer of STO/PTO/STO toroidal arrangement showed the evolution of the vortex state from the ferroelectric state at head-to-head domain boundaries of the a_1/a_2 twin structures.[5] The vortex state was stabilized by changing the number of PTO layers to

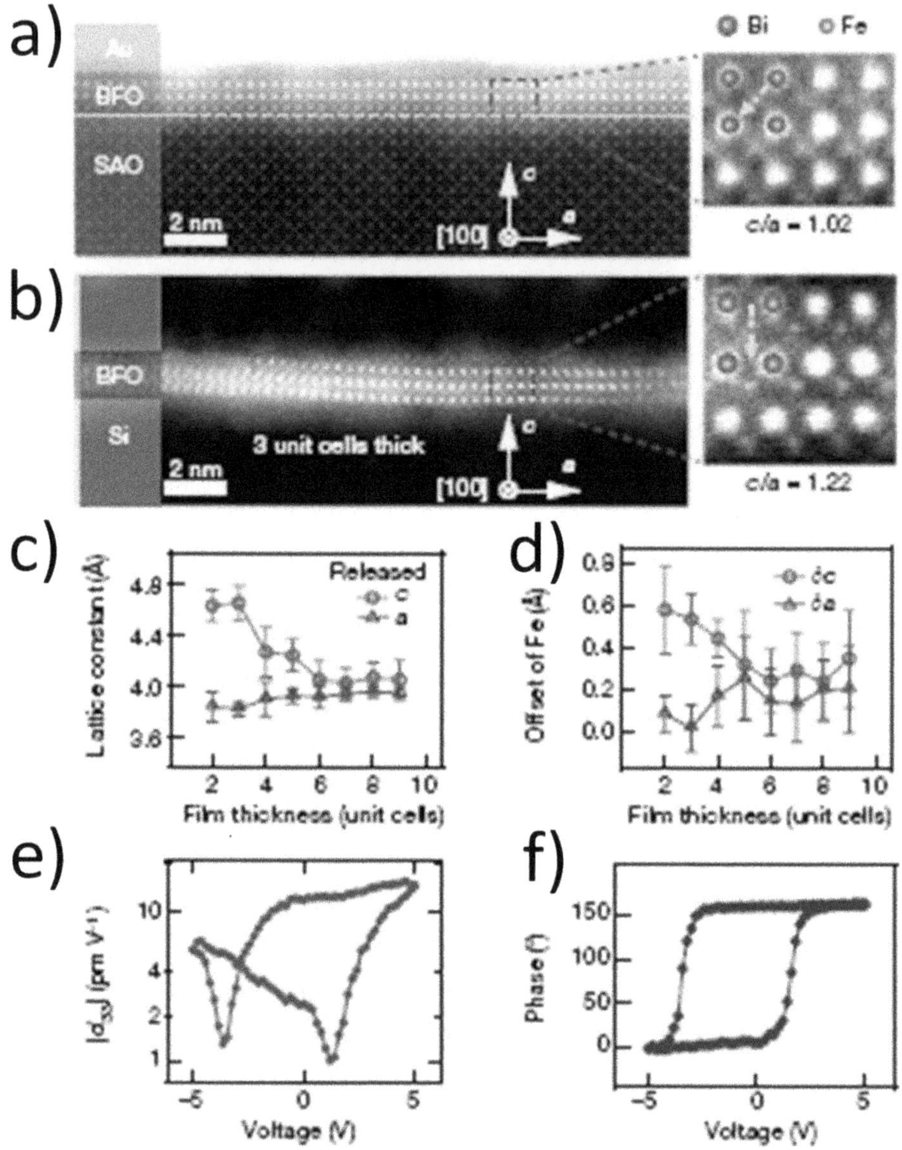

FIGURE 6.3 Giant polarization and lattice distortion in ultrathin free-standing BFO films. (**a,b**) Cross-sectional HAADF images of a three unit-cell BFO film before (**a**) and after (**b**) releasing the film, showing an R-like phase with polarization along the <111> directions and a T-like phase with polarization along the <001> direction. (**c,d**) The *c*/*a* ratio (**c**) and the offset (δc and δa) of Fe ions from the centres of four neighbouring Bi ions (**d**) as a function of the thickness of free-standing BFO films, showing the evolution from an R-like to a T-like phase transition as the thickness of the film decreases. The error bars in (**c**) and (**d**) represent the fitting error of the lattice constants. (**e,f**) PFM amplitude–voltage butterfly loop (**e**) and phase–voltage hysteresis loop (**f**) of a four-unit-cell free-standing BFO film on a conductive silicon substrate, showing that the polarization is switchable. d_{33}, out-of-plane piezoelectric coefficient. (Reproduced from Reference [3]. Creative Commons.)

deploy the long-range interactions among multiple toroidal layers in the superlattices configuration. The possibility of integrating free-standing perovskite films onto Si or other semiconducting substrates paved a new route for designing multifunctional electronic devices. Heterostructures of perovskite layers with different perovskite oxides having diverse functionalities, and 2D TMD or graphene layers are discussed in Chapter 5 in detail for their practical device applications.

REFERENCES

1. X. Zhu, Z. Liu, N. Ming, *Nanosci. Nanotech.* 10 (2010) 4109.
2. X. Zhu, Z. Liu, N. Ming, *J. Mater. Chem.* 20 (2010) 4015.
3. D. Ji, S.Cai, T. R. Paudel, H. Sun, C. Zhang, L. Han, Y. Wei, Y. Zang, M. Gu, Y. Zhang, W. Gao, H. Huyan, W. Guo, D. Wu, Z. Gu, E. Y. Tsymbal, P. Wang, Y. Nie, X. Pan, *Nature* 570 (2019) 87.
4. L. Li, L. Liang, H. Wu, X. Zhu, *Nanoscale Res. Lett.* 11 (2016) 121.
5. X.-J. Hu, Y. Yang, C. Hou, T.-X. Liang, *J. Phys. Chem. C* 126 (2022) 517.
6. J. Lu, W. Luo, J. Feng, H. Xiang, *Nano Lett.* 18 (2018) 595.
7. W. E. Teo, S. Ramkrishna, *Nanotechnol.* 17 (2006) R89.
8. H. S. Gu, Y. M. Hu, J. You, Z. L. Hu, Y. Yuan, T. J. Zhang, *J. Appl. Phys.* 101 (2007) 024319-1.
9. U. A. Joshi, S. Yoon, S. Baik, J. S. Lee, *J. Phys. Chem. B* 110 (2006) 12249.
10. Y. M. Hu, H. S. Gu, X. C. Sun, J. You, J. Wang, *Appl. Phys. Lett.* 88 (2006) 193120-1.
11. X. H. Zhu, J. Y. Wang, Z. H. Zhang, J. M. Zhu, S. H. Zhou, Z. G. Liu, N. B. Ming, *J. Am. Ceram. Soc.* 91 (2008) 2683.
12. G. Xu, Z. H. Ren, P. Y. Du, W. J. Weng, G. Shen, G. R. Han, *Adv. Mater.* 17 (2005) 907.
13. J. Wang, C. S. Sandu, E. Colla, Y. Wang, W. Ma, R. Gysel, H. J. Trodahl, N. Setter, M. Kuball, *Appl. Phys. Lett.* 90 (2007) 133107-1.
14. J. J. Urban, J. E. Spanier, L. Ouyang, W. S. Yun, H. K. Park, *Adv. Mater.* 15 (2003) 423.
15. J. J. Urban, W. S. Yun, Q. Gu, H. Park, *J. Am. Chem. Soc.* 124 (2002) 1186.
16. Y. B. Mao, S. Banerjee, S. S. Wong, *J. Am. Chem. Soc.* 125 (2003) 15718.
17. N. Z. Bao, L. M. Shen, A. Gupta, A. Tatarenko, G. Srinivasan, K. Yanagisawa, *Appl. Phys. Lett.* 94 (2009) 253109-1,
18. X. F. Lu, D. L. Zhang, Q. D. Zhao, C. Wang, W. J. Zhang, Y. Wei, *Macromol. Rapid Commun.* 27 (2005) 76.
19. X. Y. Ma, H. Zhang, J. Xu, J. J. Niu, Q. Yang, J. Sha, D. Yang, *Chem. Phys. Lett.* 363 (2002) 579.
20. K. S. Shankar, A. K. Raychaudhuri, *Nanotechnol.* 15 (2004) 1312.
21. F. Chen, H. W. Liu, K. F. Wang, H. Yu, S. Dong, X. Y. Chen, X. P. Jiang, Z. F. Ren, J. M. Liu, *J. Phys. Condens. Matter* 17 (2005) L467.
22. T. Zhang, C. G. Jin, T. Qian, X. L. Lu, J. M. Bai, X. G. Li, *J. Mater. Chem.* 14 (2004) 2787.
23. D. L. Zhu, H. Zhu, Y. H. Zhang, *J. Phys. Condens. Matter* 14 (2002) L519.
24. D. L. Zhu, H. Zhu, Y. H. Zhang, *Appl. Phys. Lett.* 80 (2002) 1634.
25. S. S. Rao, K. N. Anuradha, S. Sarangi, S. V. Bhat, *Appl. Phys. Lett.* 87 (2005) 182503-1.
26. D. Zhu, H. Zhu, Y. Zhang, *J. Cryst. Growth* 249 (2003) 172.
27. S. Datta, S. Chandra, S. Samanta, K. Das, H. Srikanth, B. Ghosh, *J. Nanomater.* 2013 (2013) 162315.
28. T. Zhang, C. G. Jin, T. Qian, X. L. Lu, J. M. Bai, X. G. Li, *J. Mater. Chem.* 14 (2004) 2787.

29. C. Li, B. Lei, Z. Luo, S. Han, Z. C. Liu, D. H. Zhang, C. W. Zhou, *Adv. Mater.* 17 (2005) 1548.
30. L. Li, H. Li, X. Zhai, C. Zeng, *Appl. Phys. Lett.* 103 (2013) 113101.
31. B. A. Hernandez, K. S. Chang, E. R. Fisher, P. K. Dohout, *Chem. Mater.* 14 (2002) 480.
32. Y. Luo, I. Szafraniak, N. D. Zakharov, V. Nagarajan, M. Steinhart, R. B. Wehrspohn, J. H. Wendorff, R. Ramesh, M. Alexe, *Appl. Phys. Lett.* 83 (2003) 440.
33. Y. Luo, I. Szafraniak, V. Nagarajan, R. B. Wehrspohn, M. Steinhart, J. H. Wendorff, N. D. Zakharov, R. Ramesh, M. Alexe, *Integr. Ferroelectr.* 59 (2003) 1513.
34. Y. B. Mao, T. J. Park, S. S.Wong, *Chem. Commun.* 46 (2005) 5721.
35. Y. Y. Chen, B. Y. Yu, J. H. Wang, R. E. Cochran, J. J. Shyue, *Inorg. Chem.* 48 (2009) 681.
36. S. Singh, S. B. Krupanidhi, *Phys. Lett. A* 367 (2007) 356.
37. X. Z. Wei, A. L. Vasiliev, N. P. Padture, *J. Mater. Res.* 20 (2005) 2140.
38. Y. Chen, W. Chen, F. Guo, M. Y. Li, W. Liu, X. Z. Zhao, *Chin. Phys. B* 18 (2009) 3922.
39. B. A. Hernandez-Sanchez, K. S. Chang, M. T. Scancella, J. L. Burris, S. Kohli, E. R. Fisher, P. K. Dorhout, *Chem. Mater.* 17 (2005) 5909.
40. J. Rockenberger, E. C. Scher, A. P. Alivisatos, *J. Am. Chem. Soc.* 121 (1999) 11595.
41. L. L. Zhao, M. Steinhart, J. Yu, U. Gosele, *J. Mater. Res.* 21 (2006) 685.
42. P. M. Rørvik, K. Tadanaga, M. Tatsumisago, T. Grande, M. A. Einarsrud, *J. Eur. Ceram. Soc.* 29 (2009) 2575.
43. L. F. Liu, T. Y. Ning, Y. Ren, Z. H. Sun, F. F. Wang, W. Y. Zhou, S. S. Xie, J. Song, S. D. Luo, D. F. Liu, J. Shen, W. J. Ma, Y. L. Zhou, *Mater. Sci. Eng. B* 149 (2008) 41.
44. W. Y. Yang, X. H. Wang, *J. Am. Ceram. Soc.* 91 (2008) 3820.
45. S. Singh, S. B. Krupanidhi, *Appl. Phys. A: Mater. Sci. Process.* 87 (2007) 27.
46. A. Nourmohammadi, M. A. Bahrevar, S. Schulze, M. Hietschold, *J. Mater. Sci.* 43 (2008) 4753.
47. J. Kim, S. A. Yang, Y. C. Choi, J. K. Han, K. O. Jeong, Y. J. Yun, D. J. Kim, S. M. Yang, D. Yoon, H. Cheong, K. S. Chang, T. W. Noh, S. D. Bu, *Nano Lett.* 8 (2008) 1813.
48. S. A. Yang, K. O. Jeong, J. Kim, Y. C. Choi, J. K. Han, S. D. Bu, *J. Korean Phys. Soc.* 51 (2007) 174.
49. M. Alexe, D. Hesse, V. Schmidt, S. Senz, H. J. Fan, M. Zacharias, U. Gosele, *Appl. Phys. Lett.* 89 (2006) 172907.
50. O. M. Zhigalina, E. D. Mishina, N. E. Sherstyuk, K. A. Vorotilov, V. A. Vasiljev, A. S. Sigov, O. I. Lebedev, Y. V. Grigoriev, M. P. Desanto, R. Barberi, T. Rasing, *Ferroelectrics* 336 (2006) 247.
51. S. Y. Xu, Y. Shi, *J. Phys. D: Appl. Phys.* 42 (2009) 085301.
52. A. Nourmohammadi, M. A. Bahrevar, M. Hietschold, *J. Alloys Compd.* 473 (2009) 467.
53. H. S. Min, S.W. Jung, J. K. Lee, *J. Ceram. Process. Res.* 7 (2006) 253.
54. A. Nourmohammadi, M. A. Bahrevar, M. Hietschold, *Mater. Lett.* 62 (2008) 3349.
55. F. D. Morrison, L. Ramsay, J. F. Scott, *J. Phys. Condens. Matter* 15 (2003) L527.
56. M. Feng, W. Wang, Y. Zhou, D. C. Jia, *J. Sol-Gel Sci. Technol.* 52 (2009) 120.
57. B. I. Seo, U. A. Shaislamov, S. W. Kim, H. K. Kim, B. Yang, S. K. Hong, *Phys. E.* 37 (2007) 274.
58. T. J. Park, Y. Mao, S. S. Wong, *Chem. Commun.* 23 (2004) 2708.
59. X. Y. Zhang, X. Zhao, C. W. Lai, J. Wang, X. G. Tang, J. Y. Dai, *Appl. Phys. Lett.* 87 (2005) 143102.
60. X. Q. Xu, T. Qian, G. Q. Zhang, T. Zhang, G. A. Li, W. Wang, X. G. Li, *Chem. Lett.* 36 (2007) 112.
61. J. Wei, D. S. Xue, Y. Xu, *Scr. Mater.* 58 (2008) 45.
62. S. Singh, S. B. Krupanidhi, *J. Nanosci. Nanotechnol.* 8 (2008) 335.

63. C. T. Sousa, A. M. L. Lopes, *J. Nanosci. Nanotechnol.* 9 (2009) 6084.
64. A. G. Leyva, P. Stoliar, M. Rosenbusch, V. Lorenzo, P. Levy, C. Albonetti, M. Cavallini, F. Biscarini, H. E. Troiani, J. Curiale, R. D. Sanchez, *J. Solid State Chem.* 177 (2004) 3949.
65. P. Levy, A. G. Leyva, H. E. Troiani, R. D. Sanchez, *Appl. Phys. Lett.* 83 (2003) 5247.
66. G. Y. Zhang, J. Chen, *J. Electrochem. Soc.* 152 (2005) A2069.
67. J. Sacanell, M. G. Bellino, D. G. Lamas, A. G. Leyva, *Phys. B* 398 (2007) 341.
68. J. Wang, A. Manivannan, N. Q. Wu, *Thin Solid Films* 517 (2008) 582.
69. Y. Mao, S. Banerjee, S. S. Wong, *Chem. Commun.* 3 (2003) 408.
70. N. P. Padture, X. Z. Wei, *J. Am. Ceram. Soc.* 86 (2003) 2215.
71. F. E. Atalay, V. Yagmur, S. Atalay, H. Kaya, S. Tari, D. Avsar, *J. Optoelectron. Adv. M.* 12 (2010) 392.
72. X. Ma, H. Zhang, J. Xu, J. Niu, Q. Yang, J. Sha, D. Yang, *Chem. Phys. Lett.* 363 (2002) 579.
73. J. Curiale, R. D. Sánchez, H. E. Troiani, C. A. Ramos, H. Pastoriza, A. G. Leyva, P. Levy, *Phys. Rev. B* 75 (2007) 224410.
74. J. Wang, A. Manivannan, N. Wu, *Thin Solid Films* 517 (2008) 582.
75. A. Carretero-Genevrier, N. Mestres, T. Puig, A. Hassini, J. Oró, A. Pomar, F. Sandiumenge, X. Obradors, E. Ferain, *Adv. Mater.* 20 (2008) 3672.
76. L. Marín, L. Morellón, P. A. Algarabel, L. A. Rodríguez, C. Magén, J. M. De Teresa, M. R. Ibarra, *Nano Lett.* 14 (2014) 423.
77. G. Singh-Bhalla, A. Biswas, A. F. Hebard, *Phys. Rev. B* 80 (2009) 144410.
78. H. J. Liu, C. H. Sow, C. K. Ong, *J. Appl. Phys.* 100 (2006) 014306
79. H. Y. Zhai, J. X. Ma, D. T. Gillaspie, X. G. Zhang, T. Z. Ward, E. W. Plummer, J. Shen, *Phys. Rev. Lett.* 97 (2006) 167201.
80. I. Pallecchi, A. Gadaleta, L. Pellegrino, G. C. Gazzadi, E. Bellingeri, A. S. Siri, D. Marré, *Phys. Rev. B* 76 (2007) 174401.
81. O. Céspedes, S. M. Watts, J. M. D. Coey, K. Dörr, M. Ziese, *Appl. Phys. Lett.* 87 (2005) 083102.
82. X. Guo, P. G. Li, X. Wang, X. L. Fu, L. M. Chen, M. Lei, W. Zheng, W. H. Tang, *J. Alloy. Compd.* 485 (2009) 802.
83. K. Hayat, M. J. Iqbal, K. Rasool, Y. Iqbal, *Chem. Phys. Lett.* 616 (2014) 126.
84. B. Jugdersuren, S. Kang, R. S. DiPietro, D. Heiman, D. McKeown, I. L. Pegg, J. Philip, *J. Appl. Phys.* 109 (2011) 016109.
85. X. Zhou, Y. Zhao, X. Cao, Y. Xue, D. Xu, L. Jiang, W. Su, *Mater. Lett.* 62 (2008) 470.
86. W. S. Yun, J. J. Urban, Q. Gu, H. K. Park, *Nano. Lett.* 2 (2002) 447.
87. K. Y. Yun, M. Noda, M. Okuyama, H. Saeki, H. Tabata, K. Saito, *J. Appl. Phys.* 96 (2004) 3399.
88. M. I. Dolz, W. Bast, D. Antonio, H. Pastoriza, J. Curiale, R. D. Sanchez, A. G. Leyva, *J. Appl. Phys.* 103 (2008) 083909.
89. L. Li, X. Zhang, L. Li, X. Zhai, C. Zeng, *Solid State Commun.* 171 (2013) 46.
90. L. Ghivelder, F. Parisi, *Phys. Rev. B* 71 (2005) 184425.
91. D. Niebieskikwiat, R. D. Sanchez, *J. Phys Condens. Matter* 24 (2012) 436001.
92. J. F. Scott, H. J. Fan, S. Kawasaki, J. Banys, M. Ivanov, A. Krotkus, J. Macutkevic, R. Blinc, V. V. Laguta, P. Cevc, J. S. Liu, A. L. Kholkin, *Nano Lett.* 8 (2008) 4404.
93. X. C. Zhou, X. S. Chen, W. Lu, *Phys. B* 369 (2005) 117.
94. A. Krotkus, R. Adomavicius, G. Molis, A. Urbanowicz, H. Eusebe, *J. Appl. Phys.* 96 (2004) 4006.
95. T. Zhao, S. B. Ogale, S. R. Shinde, R. Ramesh, R. Droopad, J. Yu, K. Eisenbeiser, J. Misewich, *Appl. Phys. Lett.* 84 (2004) 750.

96. Y. Lu, W. Li, G. Gong, G. Xiao, A. Gupta, P. Lecoeur, J. Sun, Y. Wang, V. Dravid, *Phys. Rev. B* 54 (1996) R8357.
97. M. Bowen, M. Bibes, A. Barthelemy, J. P. Contour, A. Anane, Y. Lemaitre, A. Fert, *Appl. Phys. Lett.* 82 (2003) 233.
98. Y. Arimoto, H. Ishiwara, *MRS Bull.* 29 (2004) 823.
99. J. F. Scott, F. D. Morrison, M. Miyake, P. Zubko, X. J. Lou, V. M. Kugler, S. Rios, M. Zhang, T. Tatsuta, O. Tsuji, T. J. Leedham, *J. Am. Ceram. Soc.* 88 (2005) 1691.
100. L. Peña, L. Garzón, R. Galceran, A. Pomar, B. Bozzo, Z. Konstantinovic, F. Sandiumenge, L. Balcells, C. Ocal, B. Martinez, *J. Phys. Condens. Matter* 26 (2014) 395010.
101. F. D. Morrison, Y. Luo, I. Szafraniak,V. Nagarajan, R. B. Wehrspohn, M. Steinhart, J. H. Wendorff, N. D. Zakharov, E. D. Mishina, K. A. Vorotilov, A. S. Sigov, S. Nakabayashi, M. Alexe, R. Ramesh, J. F. Scott, *Rev. Adv. Mater. Sci.* 4 (2003) 114.
102. M. Alexe, Y. Luo, I. Szafraniak, R. B. Wehrspohn, M. Steinhart, *Eur. Pat.* (2002) 03000969.0.
103. H. Y. Sun, C. C. Zhang, J. M. Song, J. H. Gu, T. W. Zhang, Y. P. Zang, Y. F. Li, Z. B. Gu, P. Wang, Y. F. Nie, *Thin Solid Films* 697 (2020) 137815.
104. L. Han, Y. Fang, Y. Zhao, Y. Zang, Z. Gu, Y. Nie, X. Pan, *Adv. Mater. Interfaces* 7 (2020) 1901604.
105. Q. Wang, H. Fang, D. Wang, J. Wang, N. Zhang, B. He, W. Lu, *Crystals* 10 (2020) 733
106. G. Dong, S. Li, T. Li, H. Wu, T. Nan, X. Wang, H. Liu, Y. Cheng, Y. Zhou, W. Qu, Y. Zhao, B. Peng, Z. Wang, Z. Hu, Z. Luo, W. Ren, S. J. Pennycook, J. Li, J. Sun, Z.-G. Ye, Z. Jiang, Z. Zhou, X. Ding, T. Min, M. Liu, *Adv. Mater.* 32 (2020) 2004477.
107. F. Zhou, Y. Li, X. Liao, S. Lin, H. Song, Y. Liu, F. Yang, S. Yan, Z. Lai, Y. Liu, F. Yang, S. Yan, Z. Lai,Y. Liu, C.-Y. Xu, Z. Yang, Y. Huang, L. Zhen, J. Zhu, J. Yao *Nano Lett.* 21 (2021) 4700.
108. S. Li , Y. L. Zhu, Y. L. Tang, Y. Liu, S. R. Zhang, Y. J. Wang, X. L. Ma, *Acta Materialia* 131 (2017) 123.
109. O. Nesterov, S. Matzen, C. Magén, A. H. G. Vlooswijk, G. Catalan, B. Noheda, *Appl. Phys. Lett.* 103 (2013) 142901.
110. S. P. Alpay, A. L. Roytburd, *J. Appl. Phys.* 83 (1998) 4714.

7 Few Special Transition Metal Oxides

1D Spin Dynamics

7.1 VO_2: 1D FINITE-SIZED HEISENBERG SPIN CHAIN

VO_2, the most popular correlated transition metal oxide, turns out to be the centre of extensive research because of its first-order, reversible metal-to-insulator transition (MIT) near room temperature (~340 K).[1] The MIT, however, is associated with a structural phase transition (SPT) from monoclinic M1 (space group $P2_1/c$) to rutile tetragonal R (space group $P4_2/mnm$).[2,3] Moreover, an optical transition from transmitting infrared (IR) at low temperature to reflecting IR at high temperature is also reported for VO_2.[4] These transitions make VO_2 a promising material for various thermal, electrical, and optical devices such as smart windows,[5] gas sensors,[6] electrical switches,[7] and cathodes for Li-ion batteries.[8] Two other low-temperature phases of VO_2 (monoclinic M2; space group $C2/m$; and triclinic T; space group $C\overline{1}$) are also reported to be present in the structural phase diagram of VO_2.[9] The structural shape of VO_2 is made of by parallel vanadium chains surrounded by oxygen atoms forming distorted octahedron. In the R phase, V chains are parallel, and V atoms are periodic along the c_R axis,[10] whereas in the M1, M2, and T phases, the V atoms arrange themselves differently along the c_R axis (Figure 7.1). The V atoms dimerize and forms pairs with a tilting along the c_R axis in the M1 phase. The unit cell volume also doubles that of the R phase (schematic in Figure 7.1b) with the new crystallographic parameters as $a_{M1} \leftrightarrow 2c_R$, $b_{M1} \leftrightarrow a_R$, and $c_{M1} \leftrightarrow b_R\text{-}c_R$.[11]

In the M2 phase, the arrangement of V atoms creates two sublattices: in one sublattice, V atoms tilt along the c_R axis but don't form pairs, while the V ions in the next sublattice form pairs, but the pairs don't twist away from the c_R axis (Figure 7.1c). The crystallographic relationship with the R phase is $b_{M2} \leftrightarrow 2c_R$, $a_{M2} \leftrightarrow 2a_R$, and $c_{M2} \leftrightarrow -b_R$.[12] The T phase resembles that of the M2 phase, with a slight change in the twisting of the V pairs, making the structure triclinic.[13] The simultaneous existence of MIT and SPT in VO_2 opens a conflict on whether strong electron-electron correlation or electron-phonon coupling initiates the MIT.[14–16] M1 is an insulating nonmagnetic phase, which suggests the transition from M1 to metallic R follows Peierls scenario.[6] However, the existence of an insulating magnetic M2 phase, where V ions in alternate chains are dimerized, indicates that only Peierls-type dimerization does not drive the MIT. Goodenough, in 1971, anticipated[17] the electronic phase diagram of VO_2 where

DOI: 10.1201/9781003623694-7

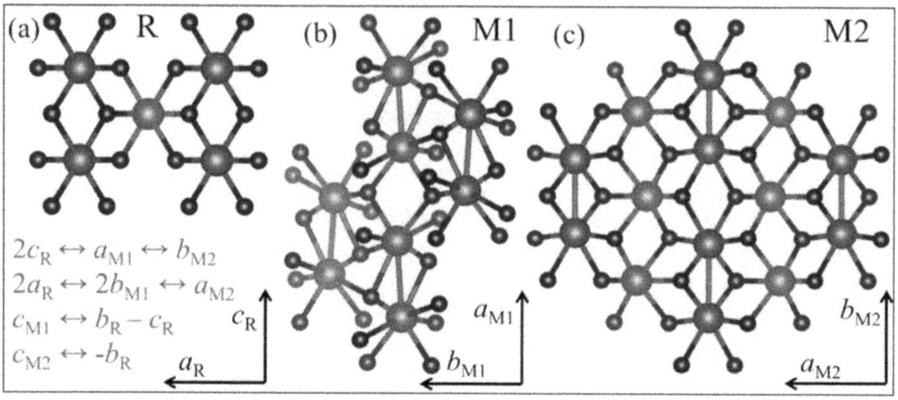

FIGURE 7.1 The schematic structures for (**a**) rutile (R), monoclinic (**b**) M1, and (**c**) M2 phases of VO_2. Large and small balls denote V and O atoms, respectively. (Reproduced from Reference [12]. Creative Commons.)

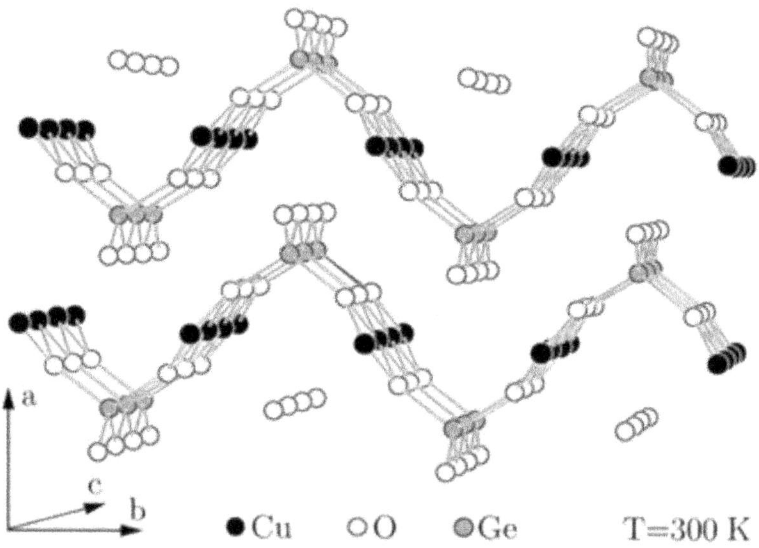

FIGURE 7.2 Schematic electronic band structures for the insulating and metallic phase of VO_2. (Reproduced from Reference [12]. Creative Commons.)

the lower lying t_{2g} states of V 3d levels split into a_{1g} and e_g^π doublet by the effect of the tetragonal crystal field (schematic in Figure 7.2). The tilting of the V-V dimers leads to two effects. First, the a_{1g} band is split into a lower-energy bonding combination and a higher-energy antibonding one. Second, the tilting of the pairs increases the overlap of these states with O states, pushing the e_g^π states to higher energy. In Goodenough's picture, the single d electron occupies the a_{1g} bonding combination (half-filled band).

In the metallic R phase, all three (a_{1g}, $a_{1g'}$, and e_g^{π}) bands are partially occupied, whereas in the insulating phase, only a_{1g} is half-filled and overlapping $a_{1g'}$ and e_g^{π} bands are empty with a gap between them of the order of 0.6 eV (Figure 7.2).

Zylbersztejn and Mott,[12] Sommers and Doniach,[13] and Rice *et al.*[14] suggested that Coulomb repulsion plays the main role in opening an energy gap at the lower temperature. However, no gap was found between a_{1g} and overlapped $a_{1g'}$, e_g^{π} bands using density functional theory (DFT) with local density approximation (LDA),[15] even including structural distortion in the calculation as demonstrated by Goodenough.[17] This shows that SPT may not induce the MIT in VO_2. However, Biermann *et al.*[18] showed the opening of a bandgap, though much smaller than the observed experimental value, using LDA along with cluster dynamical mean field theory (DMFT), which was found to be increased with local Coulomb correlation strength. The report suggested that the Coulomb correlation might be sufficient to cause separation between a_{1g} and overlapped $a_{1g'}$, e_g^{π} states, leading to an orbital selection and a half-filled a_{1g} band. The orbitals of the half-filled a_{1g} band are such that they have appreciable hopping only along the c_R-axis. So, the reduction from three-band to one-band coincides with a reduction of dimensionality from three to one. One-dimensionality of the a_{1g} band also leads to an increase in effective locally screened Coulomb repulsion energy (Hubbard *U*) and drives the chains toward Mott-Hubbard insulating phase (non-interacting, insulating Heisenberg spin ½ chains). Again, because of one-dimensionality, Heisenberg spin ½ chains are susceptible to spin-Peierls (dimerized phase) transition accompanying SPT. Basu *et al.*[19] showed that the electronic transition went successively from three-band, 3D metal to a single half-filled band, then to a 1D Heisenberg chain and finally to a spin-Peierls insulator. Reduction to 1D due to orbital selection is the reason for SPT and MIT to happen at the same time in a discontinuous fashion.

7.1.1 Growth

The conflict between the various phase transitions of VO_2 makes it popular in the scientific community. However, the phase transitions strongly depend on the stoichiometry, dimensionality, and morphology of the structure. Doping and crystallography were also known to play a crucial role in the phase transitions of VO_2.[20,21] The observation provides the necessity for investigation of the synthetic procedure, both from fundamental and applied points of view. In this section, we will discuss the technology and methods for the synthesis of low-dimensional VO_2-based structures.

The major techniques used for the synthesis of low-dimensional (thin films and nanostructures) VO_2 are physical vapour deposition (PVD), chemical vapour deposition (CVD), solgel, and hydrothermal methods. CVD is the most widespread method for obtaining high-purity thin films and nanostructures for large-scale growth. Nag and Haglund[22] discussed the application of CVD for VO_2 synthesis. The new-age techniques for CVD-grown thin films, along with the mechanism, were presented by Vernardou and Pemble[23] The deposition temperature in the CVD procedure plays a vital role. Su *et al.*[24] synthesized V_2O_5 rods or VO_2 micro-blocks, at 500 C and 150 C, respectively, on the glass surface using the same precursor vanadium (IV) acetylacetonate. Moreover, a controlled way of doping is possible using the CVD technique. W-doped[25] and F-doped[26] thin films of VO_2 are reported on the glass surface. PVD is also known as an excellent tool for the preparation of 2D VO_2 thin films. Yi *et al.*[27] reported ion-beam sputtered VO_2 film using metallic vanadium as a target on a Si wafer following post-oxidation. However, controlled oxygen pressure and optimum temperature of the substrate are needed to be followed to get stoichiometric VO_2. Aliev and Klimov[28] used laser ablation under strict control of the substrate's temperature and oxygen pressure using 99.9% metallic vanadium as a target. However, other Magnelli phases of vanadium were obtained on TiO_2 (110)[29] and SnO_2 (110)[30] substrates following similar conditions. Beydaghyan *et al.*[31] used a radio frequency (RF) magnetron sputtering technique to synthesize VO_2 films on indium tin oxide (ITO)-coated glass substrates. VO_2 film may also be synthesized by reducing a V_2O_5 film. Saitzek *et al.*[32] prepared ~100 nm thick VO_2 (B) and VO_2 (M/R) films by reducing RF-sputtered V_2O_5 films on Si (001) substrate. Pulsed laser deposition (PLD) is also an alternative technique widely used for the preparation of VO_2 thin films. Li and Dho[33] reported an orientation-dependent modification of MIT temperature in VO_2/TiO_2 hetero-epitaxial bilayers. PLD technique was also used for doping metals in VO_2 on Si[34] and quartz[35] substrates. The solgel method is also used for the synthesis of VO_2 thin films and nanostructures. In the solgel synthesis of VO_2, the initial Magnelli phases were converted into VO_2 (B) and VO_2 (M).[36] The solgel method is also used to prepare doped VO_2 thin films, with high homogeneity, on sapphire,[37] Al,[38] Si,[39] or glass[40] substrates. Doping of Fe and Cr ions into VO_2 is discussed by Nazarov *et al.*[41] using the atomic layer deposition technique, where they have shown the role of synergetic effect in reducing the MIT temperature. VO_2 nanoparticles can be synthesized by the hydrothermal method. Hydrothermal synthesis is based on the reduction of water-soluble vanadium compounds, preferably NH_4VO_3[42,43] and V_2O_5,[44,45] in the presence of reductants[46,47] at elevated temperatures and pressures.

7.1.2 Structural Phase Transition and Metal-to-Insulator Transition

As MIT and SPT occur simultaneously in VO_2, it opens a 'chicken and egg' type debate on which takes control over the phase transition. As discussed in Section 7.1, in the M1 phase, all V ions are paired into dimers that are tilted away from the c_R-axis. However, excess oxygen or, equivalently, vanadium vacancy (V_v) may lead to the formation of V^{5+} (holes in the spin ½ chains).[19,48] The holes effectively cut off the 1D Heisenberg spin chains into smaller fragments (infinite chains becoming finite in size). Basu *et al.*[19] showed that the MIT occurred in this finite-size system when the correlation length became of the order of the system size with the critical exponent value, $\nu = 2/3$. However, the value of the critical exponent suggests that the chains will undergo dimerization in this finite-size spin ½ Heisenberg chain.[49,50] Increase in the T_c value with a decrease in sample size was also reported by a few earlier workers in small particles[51] and thin films,[52] invoking the presence of strain in the system. The T_c is also reported to increase in Mg^{+2}, Cr^{+3}, Al^{+3}, Ga^{+3} ions doped VO_2.[9,20,21,53]

These systems fall in the same universality class as hole-rich materials.[19] Other than temperature, external electric field,[54] hydrostatic pressure,[55] and strain[19,56] can also introduce phase transition in VO_2. Low-frequency Raman modes (~40 to 150 cm^{-1}), as reported by Basu *et al.*,[53] represented as spin waves were observed to propagate independently from the charge density waves. The shift in the frequency of spin wave with Mg doping in the absence of any SPT confirms that the SPT does not prompt the MIT. In another study, Basu *et al.*[55] showed that the separation of V-V dimers along [100] direction was responsible for the SPT from M1 to M2 phase of VO_2 using high-temperature TERS studies on single VO_2 nanorod. Moreover, a pressure-induced SPT of M2→T→M1 confirms that a first-order Mott-type MIT initiates a second-order SPT without any latent energy, leading to SPT and MIT observed simultaneously in VO_2.

7.1.3 Magnetic Properties

The two major models to explain the phase transitions in VO_2, as discussed in Section 7.1, are Peierls (structural distortion) and Mott physics (electronic correlation). However, one more aspect to explain the phase transition of VO_2, though rarely discussed, is Slater[57] model of magnetic enhancement. The other two vanadium oxides, V_2O_5 and V_2O_3, known to be undergoing MIT are reported to show magnetic ordering via spin alignment, which could result in bandgap modification.[58,59] A magnetic measurement by Zhang *et al.*[60] suggested that VO_2 experiences Pauli and unusual paramagnetic behaviours above and below the transition temperature, respectively, as characterized by an abrupt decrease in susceptibility near the metal-insulator transition temperature. Safi *et al.*[61] reported a reduction of >80% spin pumping signal and an increase of 5 times charge-spin conversion efficiency upon insulator to metal transition in VO_2 by employing ferromagnetic resonance-driven spin pumping and the inverse spin Hall effect measurement. Moreover, a small change in stoichiometry or a small percentage of doping may induce magnetism in VO_2. Guo *et al.*[62] reported ferromagnetism in VO_2(M1) nanobelts at room temperature due to the formation of oxygen vacancies. Biswas and Raychaudhury[63] reported ferrimagnetic metallic behaviour in Cr-doped VO_2.

7.1.4 Thermoelectric Behaviour

Thermally generated resistivity modulation is a well-studied method to induce a MIT in VO_2.[19] This thermoelectric property near room temperature makes VO_2 applicable as heat-activated devices. The phase transition in VO_2 can be induced either by electric-field-mediated generation of holes and electrons or via Joule heating or electric heating by an electric current flow.[64] Recent studies show that a thermal or laser-induced phase transition in VO_2 causes band collapse by changing the electron correlation instantaneously and finally stabilizing through structural evolution.[64] Roach and Balberg[65] showed that energy conversion between photons and phonons induced a reversible phase transition of VO_2 while illuminated by an ultrafast laser. Moreover, femtosecond lasers and visible light pulses induced phase transitions show that when the laser flux increases to a certain size, VO_2 undergoes a structural transition from the insulating monoclinic phase to the metallic rutile phase.[66] Stefanovich *et al.*[67] in 2000 reported, for the first time, that applying bias and electron injection can induce a reversible phase transition of VO_2. Lee *et al.*[68] explained that a reduction in the critical voltage for triggering the phase transition is possible by increasing the temperature. A two-terminal device using VO_2 nanowires shows a MIT at a bias voltage of 0.34 V.[69]

7.1.5 Spin Waves

Observation of spin wave, for the first time, and its manifestation in the phase transition in VO_2 is established by Basu *et al.*[53] using both phenomenological model and experimental evidence along with spinon mode frequency calculation. The phase separation in VO_2 is described by the Raman mode frequencies observed at low energies. The low-frequency modes were modes associated with the collective spinon vibration (spin wave) considering soliton, anti-soliton, and breathers modes of the dimerized 1D spin ½ Heisenberg chains (Figure 7.3a).[70] Moreover, the significant deviation of Stokes to anti-Stokes intensity ratio of the low-frequency spinon mode (Figure 7.3b) from the usual thermal factor $exp(h\upsilon/K_BT)$ of Boltzmann's distribution law confirms it to originate due to spin wave. The orthogonal dependency of the phonon and spinon vibration in the polarized Raman study also confirms their claim. The shift in the

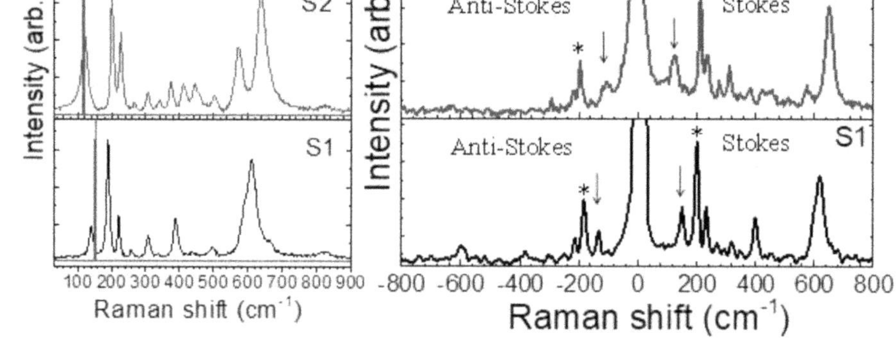

FIGURE 7.3 (**a**) Typical Raman spectra for M1 (S1), T (S2), and M2 (S3) phase of VO_2 collected at 80 K and corresponding δ-functional peak at the calculated Raman mode frequency. (**b**) Stokes and anti-Stokes spectra for the samples S1, S2, and S3. The spin-wave and the phonon mode are denoted by *arrow* and *star* signs, respectively. (Reproduced from Reference [53]. Creative Commons.)

spinon vibration followed by the increase in MIT temperature without any structural change confirms that SPT and MIT in VO_2 are governed by two separate phenomena. Moreover, the presence of spin wave confirms that perturbation due to spin-Peierls dimerization induces SPT, whereas MIT is solely driven by Mott physics. Thus, the observation of spin wave resulting from 1D spin ½ Heisenberg chain resolves the years-long debate in VO_2 and can be extended to oxide-based multiferroics, which are useful for various potential device applications.

7.1.6 Applications

7.1.6.1 Electrical Switches

VO_2 undergoes a MIT near room temperature with a change in resistance of ~three to five orders.[19,53] VO_2 finds its application as photoelectric as well as terahertz switches. Schalch *et al.*[71] designed an electrically tuned terahertz switch of VO_2 with suppressed reflection. VO_2 also finds its application as the channel in metal oxide semiconductor FET (MOSFET). However, it operates in a different fashion and is known as the Mott transition field effect transistor (MTFET).[72] MTFET shows superior efficiency than MOSFET due to its characteristic reversible phase transition and the ability of changing state from pure metallic to perfect insulating rather than only modifying the resistance. However, the threshold value of carrier density to prompt the MIT of VO_2 is ~10^{18}-10^{19} cm^{-3}, and thus could not be achieved by conventional solid gating.[73] The convenience of the ionic liquid or electrolyte as gate material in the MIT of VO_2 is more significant.[74,75] Nakano *et al.*[74] reported a SPT by gating due to huge carrier delocalization across the MIT. Moreover, surface metallicity-induced gating was explained by Liu *et al.*[76] In another study, Zhou *et al.*[77] reported a suppression of the resistance modulation by controlled gating of VO_2 films coated with a monolayer graphene. However, the incorporation of the ionic liquid as gate material may react electrochemically and alter the nature of MIT.[78] To overcome this, Singh *et al.*[79] have demonstrated irreversible phase transition in the non-stoichiometric VO_2 thin films.

7.1.6.2 Thermochromic Effect

The variation in the colour due to a change in optical properties by applying temperature is called the thermochromic effect.[80] VO_2 exhibits an optical transition as a function of temperature,[81] and therefore acts as a thermochromic material and can be used in smart window applications.[82,83] VO_2 is almost transparent to the IR range in the insulating phase but reflects in the metallic phase.[81,82] Thin films of VO_2 can be coated on the glass window to utilize solar radiation more efficiently. Such a smart window is best suited for controlling the surroundings of a closure especially in the Middle East countries, where the temperature is generally high. However, to enhance the applicability of VO_2 as a smart-window coating, a doping-induced reduction in the transition temperature toward room temperature has been studied. Doping of W in VO_2 shows a significant reduction in transition temperature.[84] Moreover, various buffer layers were utilized to improve the thermochromic properties of VO_2, for example, ZnO, SnO_2, TiO_2, and Al_2O_3.[85-87] In addition, to make the coating anti-reflecting, and antifogging, multilayers and porous structures of VO_2 have been prepared.[88,89]

7.1.6.3 Sensors

VO_2 changes its electrical, optical, and thermal properties as a function of temperature, pressure, electric and magnetic field, and therefore can be applicable to various sensors, such as biosensors, pressure sensors, temperature sensors, gas sensors, and so on.[6,90] Kim *et al.*[91] prepared temperature sensors of VO_2 films on an Al_2O_3 substrate that can be used for programming. The sensor can measure the change of conductivity by applying a voltage to the VO_2 film through the Ni electrode. The applied external voltage and corresponding bidirectional stresses can regulate the detectable temperature range. Azharudeen *et al.*[92] prepared a glucose sensor using mesoporous VO_2/PVA nanocomposite with high sensitivity, selectivity, and stability. The sensing response shows a relative error of less than 4% compared with the clinical blood glucose analyser. Prasad *et al.*[6] and Basu *et al.*[90] reported that the sensing response of VO_2 toward CH_4 is higher for the semiconducting state than that of a metallic one as the relative change of resistance decreases in the later state.

7.1.6.4 Other Applications

There are several other applications for VO_2 at its lower dimension. Kats *et al.*[93] reported that VO_2 has a large negative differential thermal emittance across the MIT. A stress-driven alteration of conductance makes VO_2 applicable as a stress sensor.[94] The MIT of VO_2 makes it applicable as micro-thermometers at a sharp temperature point, which can be used for quantitative evaluation of electron beam heating, light absorption, and heat transfer of other micro-objects.[95,96] Xiao *et al.*[97] established an infrared camouflage device based on a layer of VO_2 supported by a graphene-CNT film. In the field of energy storage, Yoon *et al.*[98] reported hydrogen storage applications using VO_2 thin films. Basu *et al.*[99] reported binder-free supercapacitor performance of porous VO_2 nanostructures with capacitance of 33 $mF \times cm^{-2}$, Coulombic efficiency of 98.5%, and capacitance retention of 94.9% after 2000 cycles. The thermal sensitivity helps VO_2 to make high-quality, real-time infrared imaging from the detected object.[100]

7.2 CUGEO$_3$: 1D LINEAR SPIN-PEIERLS DIMERS

$CuGeO_3$ is the first-known inorganic spin-Peierls material and thus attracts considerable attention for both physical study and practical applications. The spin-Peierls transition in $CuGeO_3$ was discovered by Hase *et al.*[101] at 14 K. The cuprate $CuGeO_3$ crystallizes in the orthorhombic[102,103] (space group: P_{bmm}) structure with two formula units per unit cell related by the *b* glide plane (Figure 7.4). $CuGeO_3$ is composed of anisotropic AFM Cu-O and Ge-O chains along the *c* direction. The AF chains consist of S = $1/2Cu^{2+}$ ions positioned at the centre of the edge, sharing squares of oxygen atoms. Moreover, these chains are separated by rigid chains of GeO_4 tetrahedra. The structure is unstable toward a lattice dimerization and opens a spin-Peierls gap in the magnetic spectrum below 14 K.[104] This opening of bandgap includes a magnetoelastic coupling between the 1D electronic system and the 3D phonon system.[105] $CuGeO_3$ exhibits excellent electrochemical, magnetic, and optical properties and

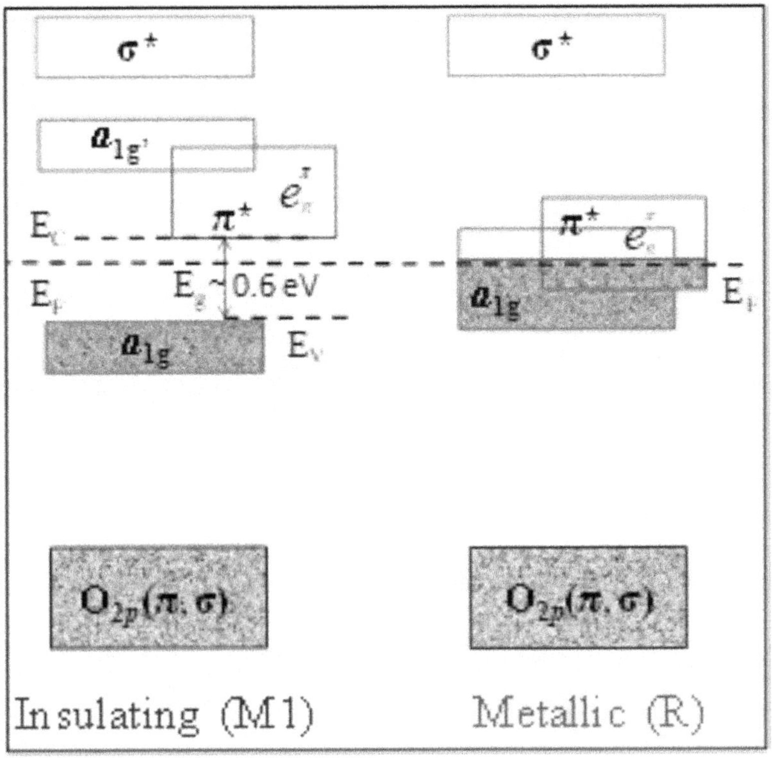

FIGURE 7.4 Schematic structure of $CuGeO_3$ at 300 K. (Reprinted with permission from Reference [102]. Copyright 2020, American Physical Society.)

therefore finds applications in nanoscale electrochemical, magnetic, and optical devices and sensors.[106-109]

7.2.1 Growth

The 1D spin-Peierls transition is one of the most attractive phenomena, especially for low-dimensional quantum spin systems. Therefore, the synthesis of low-dimensional $CuGeO_3$ nanostructures finds tremendous attention. Conventional hydrothermal synthesis is the most popular method to obtain $CuGeO_3$ in the form of 1D nanostructures. Pei *et al.*[110] used a facile hydrothermal deposition process to synthesize single crystalline $CuGeO_3$ nanowires at high temperatures. Flower-like $CuGeO_3$ nanostructures were also prepared by the same group by controlling the compactness in a simple hydrothermal route.[111] Song *et al.*[106] synthesized single-crystal $CuGeO_3$ nanobelts by a surfactant-assisted hydrothermal process. $CuGeO_3$ nanorods were prepared by Feng *et al.*[112] for application as lithium-ion batteries (LIBs). Wang *et al.*[113] synthesized single-crystalline $CuGeO_3$ nanorods of diameter ~20-35 nm with a length up to 1 μm by a facile hydrothermal route for photocatalyst application in organic pollutant degradation under UV light irradiation. Borhade *et al.*[114] prepared $CuGeO_3$ nanocomposites by mechanochemical solid state synthesis method with a green chemistry approach for photocatalyst applications. Bandgap-modulated $CuGeO_3$ nanocrystals were synthesized by Suzuki *et al.*[115] via microwave-assisted hydrothermal method and used for photocatalytic applications.

7.2.2 Structural Phase Transition under Pressure

$CuGeO_3$ is the first inorganic compound to undergo a spin-Peierls transition at 14 K while cooled down at ambient pressure.[101] $CuGeO_3$ also undergoes pressure-dependent phase transition sequences, depending on the nature of the pressure medium.[116-119] An unusual reversible phase transition was observed at a pure hydrostatic pressure range of 7 GPa with either 4:1 methanol:ethanol mixture or helium gas as a pressure-transmitting medium.[116,117] The phase transition occurs due to a sudden huge contraction (~20%) of the single-crystal platelets along the *b*-direction, and a change in colour from light blue to pale green is also observed. However, with a further increase in pressure to 14–18 GPa, the pale-green phase amorphized, as observed by Raman scattering studies.[116,120] However, with other solid (NaCl, KBr, CsI), gas (Ar, Xe), and liquid (water, pure ethanol, and petroleum) pressure-transmitting medium, the transition is suppressed, and a different transition sequence occurs. The orthorhombic phase of $CuGeO_3$ transforms the colour to blue near 7 GPa and lasts for only <1 GPa and then transforms its colour to green near 8 GPa and persists up to 20 GPa.[121] However, at higher pressures than 20 GPa, the Raman fingerprints change, indicating a further pressure-induced transition.

Adams *et al.*[122] reported the phase transition near 7 GPa as a structural transition from orthorhombic to monoclinic phase by pressure-dependent X-ray diffraction study. They also claimed a further structural transition to a higher symmetric phase at 11 and 21 GPa. Bräuninger *et al.*[123] also investigated a monoclinic phase transition under truly hydrostatic pressure as suggested by Adams *et al.*[122] However, Bräuninger *et al.*[123] reported a much lower bulk modulus (K_0 = 39.5 GPa) for the orthorhombic phase compared with the value reported by Adams *et al.*[122] (K_0 = 66 GPa). Ming *et al.*[124] carried out a pressure-induced phase transition sequence of up to 42 GPa in a diamond-anvil cell under synchrotron radiation. They observed the transition as orthorhombic phase (I) → orthorhombic phase (II) (7 GPa) → monoclinic distorted phase (20.5 GPa) → tetragonal garnet-like phase (32 GPa) while using 4:1 methanol:ethanol mixture as the pressure medium. However, with solid NaCl as the pressure medium, the phase-transformation sequence is orthorhombic phase (I) to a monoclinic phase at 8 GPa, which further transforms to an orthorhombic phase at 18.4 GPa.[124]

7.2.3 Magnetic Properties

$CuGeO_3$ is an established spin-Peierls system where a magnetic quantum fluctuation induces lattice distortions. The dynamical magnetoelastic properties predict it as a

1D Heisenberg antiferromagnetic spin ½ system.[125,126] The dimerization below the transition temperature leads to a structural transition toward lower crystal symmetry but without any long-range magnetic order.[127] However, $CuGeO_3$ follows short-range magnetic correlations due to the concomitant effect of low dimensionality and geometrical frustration.[128] Green *et al.*[129] show that below 10 K, the magnetic susceptibility fits very well to the Bulaevskii model with the mean exchange constant (*J*) of 88 K, and a ratio of 0.69 between the Cu-Cu exchange constants in the dimerized chain. Hase *et al.*[130] studied the influence of doping on the magnetic properties of $CuGeO_3$, where they observed that the effects of substitution of Mg were weaker than those of Zn. A 6% substitution of Cu by Mg lowers the spin-Peierls transition temperature and the critical field associated with the transition. However, a spin-glass-like transition was not observed, unlike with Zn doping.

7.2.4 Thermoelectric Behaviour

The thermal conductivity of the spin-Peierls compound $CuGeO_3$ was measured by Ando *et al.*[131] in the presence of magnetic fields up to 16 T. Their observation revealed that above the transition, the heat transport was mostly dominated by spin excitations, while below the transition, the spin excitations rapidly diminished. The heat transport below the transition is mostly governed by phonons. However, the phonons got scattered both by defects and by spin excitations.[131] Takeya *et al.*[132] performed the thermal conductivity measurements on both pure and Mg-doped $CuGeO_3$ in the presence of magnetic fields up to 16 T. They have observed a high-field incommensurate phase of the spin-Peierls system, the 'spin-soliton' phase in the Mg-doped $CuGeO_3$ samples. Salce *et al.*[133] measured the thermal conductivity of pure and Si-doped $CuGeO_3$ single crystals in the temperature range of 2–40 K and suggested that the transport is dominated by both phonons and pseudo-fermions.

7.2.5 Spin Waves

The dimerized spin systems consist of two spin states: (i) S = 0 singlets in the ground state, where the neighbouring spins are frozen, and (ii) S = 1 triplet, the elementary spin excitations, called a triplon. A triplon can carry spin angular momentum, known as triplon current. Chen *et al.*[134] studied the triplon current in the spin-Peierls system $CuGeO_3$. They observed that the sign of the thermal-driven triplon spin current is positive, unlike the spin wave cases in magnets. The Raman spectra of the 1D magnet $CuGeO_3$ show a broad inelastic four-spinon continuum at low temperatures below the spin-Peierls transition, which is characterized by 1D spin chains with spinon excitations (spin wave). However, at high temperatures, the intensity develops a quasi-elastic line that is analogous to the experiment.[135,136]

7.2.6 Applications

$CuGeO_3$ shows potential application for energy storage and, therefore, can be used as LIBs.[137,138] $CuGeO_3$ nanowires prepared by a facile hydrothermal reaction[139] show a reversible capacity of 690 $mAh{\cdot}g^{-1}$ at the current density of 150 $mA{\cdot}g^{-1}$ after 50 cycles. A $CuGeO_3$/rGO composite shows a reversible charge capacity of 909 mAh·g^{-1} with a high Coulombic efficiency of 91.49% at the current density of 100 mA·g^{-1} after 200 cycles.[140]

The photocatalytic activity of $CuGeO_3$ was also studied by modifying its bandgap. Borhade *et al.*[114] synthesized orthorhombic perovskite $CuGeO_3$ with a bandgap of 3.7 eV and investigated the photocatalytic activity by photochemical bleaching of Direct black dye under sunlight. Suzuki *et al.*[115] modify the bandgap of $CuGeO_3$ to a direct bandgap of ~2.44-2.91 eV by microwave-assisted hydrothermal method. They observed degradation of 50% of the methylene blue solution under UV irradiation in an interval of about 60 min.

REFERENCES

1. F. J. Morin, *Phys. Rev. Lett.* 3 (1959) 34.
2. M. W. Haverkort, Z. Hu, A. Tanaka, W. Reichelt, S. V. Streltsov, M. A. Korotin, V. I. Anisimov, H. H. Hsieh, H.-J. Lin, C. T. Chen, D. I. Khomskii, L. H. Tjeng, *Phys. Rev. Lett.* 95 (2005) 196404.
3. H.-T. Kim, Y. W. Lee, B.-J. Kim, B.-G. Chae, S. J. Yun, K.-Y. Kang, K.-J. Han, K.-J. Yee, Y.-S. Lim, *Phys. Rev. Lett.* 97 (2006) 266401.
4. E. Strelcov, A. Tselev, I. Ivanov, J. D. Budai, J. Zhang, J. Z. Tischler, I. Kravchenko, S. V. Kalinin, A. Kolmakov, *Nano Lett.* 12 (2012) 6198.
5. N. Mlyuka, G. A. Niklasson, C.-G. Granqvist, *Appl. Phys. Lett.* 95 (2009) 171909.
6. A. K. Prasad, S. Amirthapandian, S. Dhara, S. Dash, N. Murali, A. K. Tyagi, *Sens. Actuators, B* 191 (2014) 252.
7. B. Rajeswaran, A. M. Umarji, *AIP Adv.* 6 (2016) 035215.
8. C. Nethravathi, C. R. Rajamathi, M. Rajamathi, U. K. Gautam. X. Wang, D. Golberg, Y. Bando, *ACS Appl. Mater. Interfaces* 5 (2013) 2708.
9. M. Marezio, P. D. Dernier, *Acta Cryst.* A29 (1973) 618.
10. J. P. Pouget, H. Launois, J. P. D'Haenens, P. Merenda, T. M. Rice, *Phys, Rev. Lett.* 35 (1975) 873.
11. J. H. Park, J. M. Coy, T. S. Kasirga, C. Huang, Z. Fei, S. Hunter, D. H. Cobden, *Nature* 500 (2013) 431.
12. A. Zylbersztejn, N. F. Mott, *Phys. Rev. B* 11 (1975) 4383.
13. C. Sommers, S. Doniach, *Solid State Commun.* 28 (1978) 133.
14. T. M. Rice, H. Launois, J. P. Pouget, *Phys. Rev. Lett.* 73 (1994) 3042.
15. V. Eyert, *Ann. Phys. (Leipzig)* 11 (2002) 650.
16. R. Basu, *Mechanism of phase transition in vanadium oxides and its applications*, (Doc. no. PHYS 02 2015 04 013), Doctoral dissertation, Homi Bhabha National Institute (www.hbni.ac.in/phdthesis/phys/PHYS02201504013.pdf).
17. J. B. Goodenough, *J. Solid State Chem.* 3 (1971) 490.
18. S. Biermann, A. Poteryaev, A. I. Lichtenstein, A. Georges, *Phys. Rev. Lett.* 94 (2005) 026404.
19. R. Basu, M. Sardar, S. Bera, P. Magudapathy, S. Dhara, *Nanoscale* 9 (2017) 6537.
20. G. Villeneuve, M. Drillon, P. Hagenmuller, M. Nygren, J. P. Pouget, F. Carmona, P. Delhaes, *J. Phys. C: Solid State Phys.* 10 (1977) 3621.
21. W. Brückner, U. Gerlach, W. Moldenhauer, H. P. Brückner, N. Mattern, H. Oppermann, E. Wolf, *Phys. Status Solidi A* 38 (1976) 93.
22. J. Nag, R. F. Haglund Jr., *J. Phys. Condens. Matter* 20 (2008) 264016.

23. D. Vernardou, M. E. Pemble, *Thin Solid Films* 515 (2007) 8768.
24. Q. Su, C. K. Huang, Y. Wang, *J. Alloys Compd.* 475 (2009) 518.
25. C. S. Blackman, C. Piccirillo, R. Binions, I. P. Parkin, *Thin Solid Films* 517 (2009) 4565.
26. P. Kiri, M. Warwick, I. Ridley, R. Binions, *Thin Solid Films* 520 (2011) 1363.
27. X. Yi, C. Chen, L. Liu, *Infrared Phys. Techn.* 44 (2003) 137.
28. R. A. Aliev, V. A. Klimov, *Solid State Phys.* 46 (2004) 515.
29. M. D. Negra, M. Sambi, *Surf. Sci.* 494 (2001) 213.
30. A. Atrei, T. Cecconi, B. Cortigiani, U. Bardi, M. Torrini, G. Rovida, *Surf. Sci.* 513 (2002) 149.
31. G. Beydaghyan, V. Basque, P. V. Ashrit, *Thin Solid Films* 522 (2012) 204.
32. S. Saitzek, F. Guinneton, G. Guirleo, *Thin Solid Films* 516 (2008) 891.
33. J. Li, J. Dho, *J. Crystal Growth* 312 (2010) 3287.
34. B. Chen, D. Yang, P. A. Charpentier, M. Zeman, *Sol. Energy Mater. Sol. Cells* 93 (2009) 1550.
35. A. Kaushal, N. Choudhary, N. Kaur, D. Kaur, *Appl. Surf. Sci.* 257 (2011) 8937.
36. Y. Ningyi, L. Jinhua, L. Chenglu, *Appl. Surf. Sci.* 191 (2002) 176.
37. B. G. Chae, H. T. Kim, *Physica B* 405 (2010) 663.
38. J.-H. Cho, Y.-J. Byun, *Ceram. Int.* 38 (2012) S589.
39. M. Pan, H. Zhong, S. Wang, *J. Cryst. Growth* 265 (2004) 121.
40. T. J. Hanlon, J. A. Coath, M. A. Richardson, *Thin Solid Films* 436 (2003) 269.
41. D. V. Nazarov, O. M. Osmolovskaya, N. P. Bobrysheva, V. M. Smirnov, I. V. Murin, *Nanotechnol Russ.* 7 (2012) 641.
42. K. C. Kam, A. K. Cheetham, *Mater. Res. Bull.* 41 (2006) 1015.
43. W. Jiang, J. Ni, *Appl. Surf. Sci.* 257 (2011) 3253.
44. F. Sediri, F. Touati, N. Gharbi, *Mater. Sci. Eng. B* 129 (2006) 251.
45. N. Ganganagappa, A. Siddaramanna, *Mater. Charact.* 68 (2012) 58.
46. R. Li, C.-Y. Liu, *Mater. Res. Bull.* 45 (2010) 688.
47. X. Li, X. Chen, *J. Crystal Growth* 309 (2007) 43.
48. C. Chen, Z. Fan, *Appl. Phys. Lett.* 95 (2009) 262106.
49. J. Solyom, *Phys. Rev. B* 37 (1988) 5781.
50. G. Spronken, B. Fourcade, *Phys. Rev. B* 33 (1986) 1886.
51. M. Nazari, Y. Zhao, V. Hallum, A. A. Bernussi, Z. Y. Fan, M. Holtz, *Appl. Phys. Lett.* 103 (2013) 043108.
52. H. K. Kim, H. You, R. P. Chiarello, H. L. M. Chang, T. J. Zhang, D. J. Lam, *Phys. Rev. B* 47 (1993) 12900.
53. R. Basu, V. Srihari, S. K. Srivastava, S. Bera, S. Dhara, *Sci. Rep.* 10 (2020) 1977.
54. H.-T. Kim, B.-G. Chae, D.-H. Youn, G. Kim, K.-Y. Kang, *Appl. Phys. Lett.* 86 (2005) 242101.
55. R. Basu, V. Srihari, T. R. Ravindran, S. Chandra, H. K. Poswal, S. Dhara, *Phys. Rev. B* 109 (2024) 184107.
56. R. Basu, A. Patsha, S. Chandra, S. Amirthapandian, K. G. Raghavendra, A. Dasgupta, S. Dhara, *J. Phys. Chem. C* 123 (2019) 11189.
57. J. C. Slater, *Phys. Rev.* 82 (1951) 4.
58. R.-P. Blum, H. Niehus, C. Hucho, R. Fortrie, M. V. Ganduglia-Pirovano, J. Sauer, S. Shaikhutdinov, H.-J. Freund, *Phys. Rev. Lett.* 99 (2007) 226103.
59. J. Trastoy, A. Camjayi, J. del Valle, Y. Kalcheim, J.-P. Crocombette, D. A. Gilbert, J. A. Borchers, J. E. Villegas, D. Ravelosona, M. J. Rozenberg, I. K. Schuller, *Phys. Rev. B* 101 (2020) 245109.
60. R. Zhang, Q. S. Fu, C. Y. Yin, C. L. Li, X. H. Chen, G. Y. Qian, C. L. Lu, S. L. Yuan, X. J. Zhao, H. Z. Tao, *Sci. Rep.* 8 (2018) 17093.

61. T. S. Safi, P. Zhang, Y. Fan, Z. Guo, J. Han, E. R. Rosenberg, C. Ross, Y. Tserkovnyak, L. Liu, *Nat. Commun.* 11 (2020) 476.
62. D. Guo, C. Hu, Q. Yang, H. Hua, W. Li, C. Kong, *Mater. Res. Bull.* 53 (2014) 102.
63. S. Biswas, M. D. Raychaudhury, *J. Magn. Magn. Mater.* 569 (2023) 170449.
64. W. Chong, F. Liquan, L. Zhaohui, B. Jinglian, W. Shaoyan, G. Xingxing, W. Jian, Y. Wenqing, *Front. Mater.* 11 (2024) 1341518.
65. W. R. Roach, I. Balberg, *Solid State Commun.* 9 (1971) 551.
66. A. Cavalleri, Cs. Tóth, C. W. Siders, J. A. Squier, F. Ráksi, P. Forget, J. C. Kieffer, *Phys. Rev. Lett.* 87 (2001) 237401.
67. G. Stefanovich, A. Pergament, D. Stefanovich, *J. Phys. Condens. Matter* 12 (2000) 8837.
68. S. Lee, K. Hippalgaonkar, F. Yang, J. Hong, C. Ko, J. Suh, K. Liu, K.Wang, J. J. Urban, X. Zhang, C. Dames, S. A. Hartnoll, O. Delaire, J. Wu, *Science* 355 (2017) 371.
69. S.-H. Bae, S. Lee, H. Koo, L. Lin, B. H. Jo, C. Park, Z. L. Wang, *Adv. Mater.* 25 (2013) 5098.
70. M. Sato, H. Katsura, N. Nagaosa, *Phys. Rev. Lett.* 108 (2012) 237401.
71. J. S. Schalch, Y. Chi, Y. He, Y. Tang, X. Zhao, X. Zhang, Q. Wen, R. D. Averitt, *Microw. Opt. Technol. Lett.* 62 (2020) 2782.
72. S. Hormoz, S. Ramanathan, *Solid State Electron.* 54 (2010) 654.
73. S. Sengupta, K. Wang, K. Liu, A. K. Bhat, S. Dhara, J. Wu, M. M. Deshmukh, *Appl. Phys. Lett.* 99 (2011) 062114.
74. M. Nakano, K. Shibuya, D. Okuyama, T. Hatano, S. Ono, M. Kawasaki, Y. Iwasa, Y. Tokura, *Nature* 487 (2012) 459.
75. Y. J. Cui, S. Ramanathan, *J. Vac. Sci. Technol. A* 29 (2011) 041502.
76. K. Liu, D. Fu, J. Cao, J. Suh, K. X. Wang, C. Cheng, D. F. Ogletree, H. Guo, S. Sengupta, A. Khan, C. W. Yeung, S. Salahuddin, M. M. Deshmukh, J. Wu, *Nano Lett.* 12 (2012) 6272.
77. Y. Zhou J. Park, J. Shi, M. Chhowalla, H. Park, D. A. Weitz, S. Ramanathan, *Nano Lett.* 15 (2015) 1627.
78. J. Jeong, N. Aetukuri, T. Graf, T. D. Schladt, M.G. Samant, S. S. P. Parkin, *Science* 339 (2013) 1402.
79. V. R. Singh, V. Jovic, I. Valmianski, J. G. Ramirez, B. Lamoureux, I. K. Schuller, K. E. Smith, *Appl. Phys. Lett.* 111 (2017) 241605.
80. Y. Gao, H. Luo, Z. Zhang, L. Kang, Z. Chen, J. Du, M. Kanehira, C. Cao, *Nano Energy* 1 (2012) 221.
81. R. Basu, P. Magudapathy, M. Sardar, R. Pandian, S. Dhara, *J. Phys. D: Appl. Phys.* 50 (2017) 465602.
82. Y. Cui, Y. Ke, C. Liu, Z. Chen, N. Wang, L. Zhang, Y. Zhou, S. Wang, Y. Gao, Y. Long, *Joule* 2 (2018) 1707.
83. J. Schläefer, C. Sol, T. Li, D. Malarde, M. Portnoi, T. J. Macdonald, S. K. Laney, M. J. Powell, I. Top, I. P. Parkin, I. Papakonstantinou, *Sol. Energ. Mat. Sol. Cell.* 200 (2019) 109944.
84. J. Li, W. Liu, X. Zhang, P. K. Chu, K. M. C. Cheung, K. W. K. Yeung, *Mater. Today* 22 (2019) 35.
85. H. Koo, L. Xu, K.-E. Ko, S. Ahn, S.-H. Chang, C. Park, *J. Mater. Eng. Perform.* 22 (2013) 3967.
86. Z. Zhang, Y. Gao, L. Kang, J. Du, H. Luo, *J. Phys. Chem. C* 114 (2010) 22214.
87. Y. Xiong, Q.-Y. Wen, Z. Chen, W. Tian, T.-L. Wen, Y.-L. Jing, Q-H. Yang, H.-W. Zhang, *J. Phys. D-Appl. Phys.* 47 (2014) 455304.
88. C. Liu, N. Wang, Y. Long, *Appl. Surf. Sci.* 283 (2013) 222.

89. J. Y. Zheng, S. H. Bao, P. Jin, *Nano Energy* 11 (2015) 136.
90. R. Basu, P. R. Reshma, A. K. Prasad, S. Dhara, *Mat. Chem. Phys.* 248 (2020) 122901.
91. B.-J. Kim, Y. W. Lee, B.-G. Chae, S. J. Yun, S.-Y. Oh, H.-T. Kim, Y.-S. Lim, *Appl. Phys. Lett.* 90 (2007) 023515.
92. A. M. Azharudeen, R. Karthiga, M. Rajarajan, A. Suganthi, *Surf. Interf.* 16 (2019) 164.
93. M. A. Kats, R. Blanchard, S. Zhang, P. Genevet, C. Ko, S. Ramanathan, F. Capasso, *Phys. Rev. X* 3 (2013) 041004.
94. J. Cao, E. Ertekin, V. Srinivasan, W. Fan, S. Huang, H. Zheng, J. W. L. Yim, D. R. Khanal, D. F. Ogletree, J. C. Grossman, J. Wu, *Nat. Nanotechnol.* 4 (2009) 732.
95. H. Guo, M. I. Khan, C. Cheng, W. Fan, C. Dames, J. Wu, A. M. Minor, *Nat. Commun.* 5 (2014) 4986.
96. C. Cheng, D. Fu, K. Liu, H. Guo, S. Xu, S.-G. Ryu, O. Ho, J. Zhou, W. Fan, W. Bao, M. Salmeron, N. Wang, C. P. Grigoropoulos, J. Wu, *Adv. Opt. Mater.* 3 (2015) 336.
97. L. Xiao, H. Ma, J. Liu, W. Zhao, Y. Jia, Q. Zhao, K. Liu, Y. Wu, Y. Wei, S. Fan, K. Jiang, *Nano Lett.* 15 (2015) 8365.
98. H. Yoon, M. Choi, T.-W. Lim, H. Kwon, K. Ihm, J. K. Kim, S.-Y. Choi, J. Son, *Nat. Mater.* 15 (2016) 1113.
99. R. Basu, S. Ghosh, S. Bera, A. Das, S. Dhara, *Sci. Rep.* 9 (2019) 4621.
100. S. Chen, H. Ma, J. Dai, X. Yi, *Appl. Phys. Lett.* 90 (2007) 101117.
101. M. Hase, I. Terasaki, K. Uchinokura, M. Tokunaga, N. Miura, H. Obara, *Phys. Rev. B* 48 (1993) 9616.
102. A. Damascelli, D. van der Marel, G. Dhalenne, A. Revcolevschi, *Phys. Rev. B* 61 (2020) 12063.
103. H. Vollenkle, A. Wittmann, H. Nowotny, *Monatsh. Chem.* 98 (1967) 1352.
104. M. Hase, I. Terasaki, K. Uchinokura, *Phys. Rev. Lett.* 70 (1993) 3651.
105. M. Hase, *Physica B Condens. Matter.* 237 (1997) 123.
106. R. Q. Song, A. W. Xu, S. H. Yu, *J. Am. Chem. Soc.* 129 (2007) 4152.
107. L. Z. Pei, H. S. Zhao, W. Tan, H. Y. Yu, Y. W. Chen, Q. F. Zhang, C. G. Fan, *Mater. Charact.* 60 (2009) 1602.
108. Y. P. Dong, L. Z. Pei, X. F. Chu, W. B. Zhang, Q. F. Zhang, *Electrochim. Acta* 55 (2010) 5135.
109. L. Z. Pei, Y. K. Xie, Z. Y. Cai, Y. Yang, Y. Q. Pei, C. G. Fan, D. G. Fu, *J. Electrochem. Soc.* 159 (2012) K55.
110. L. Z. Pei, H. S. Zhao, W. Tan, H. Y. Yu, Y. W. Chen, Q. F. Zhang, C. G. Fan, *Cryst. Eng. Comm.* 11 (2009) 1696.
111. L. Z. Pei, Z. Y. Cai, Y. Yang, Y. Q. Pei, *e-J. Surf. Sci. Nanotech.* 9 (2011) 326.
112. J. K. Feng, M. O. Lai, L. Lu, *Mater. Res. Bull.* 47 (2012) 1693.
113. F. Wang, Y. Xing, Z. Su, S. Song, *Mater. Res. Bull.* 48 (2013) 2654.
114. A. V. Borhade, V. B. Gaikwad, Y. R. Baste, D. R. Tope, *Carbon Sci. Tech.* 5 (2013) 192.
115. V. Y. Suzuki, L. H. C. Amorin, N. H. de Paula, A. R. Albuquerque, M. S. Li, J. R. Sambrano, E. Longo, F. A. La Porta, *Mater. Today Commun.* 26 (2021) 101701.
116. A. Jayaraman, S. Y. Wang, L. C. Ming, S.-W. Cheong, *Phys. Rev.* 75 (1995) 2345.
117. A. Jayaraman, S. K. Sharma, S. Y. Wang, S.-W. Cheong, *Curr. Sci.* 71 (1996) 306.
118. A. R. Goni, T. Zou, U. Schwarz, R. K. Kremer, K. Syassen, *Phys. Rev. Lett.* 77 (1996) 1079.
119. A. Jayaraman, S. K. Sharma, S. Y. Wang, S. W. Cheong, *Phys. Rev. B* 55 (1997) 5694.
120. T. Zou, A. R. Goni, S. Ves, R. K. Kremer, K. Syassen, Phase Transitions of $CuGeO_3$ at High Pressures, *High Pressure Science and Technology* (*ed.* W. A. Treciakowski), World Scientific, River Edge, NJ, (1996) 423.

121. A. Jayaraman, S. Sharma, *Curr. Sci.* 74 (1998) 308.
122. D. M. Adams, J. Haines, S. Leonard, *J. Phys.* 3 (1991) 5183.
123. S. Bräuninger, U. Schwarz, M. Hanfland, T. Zou, R. K. Kremer, K. Syassen, *Phys. Rev. B* 56 (1997) 357.
124. L. C. Ming, S. R. Shieh, A. Jayaraman, S. K. Sharma, Y. H. Kim, *J. Phys. Chem. Solids.* 60 (1999) 69.
125. P. Pincus, *Sofia State Commun.* 9 (1971) 1971.
126. E. Pytte, *Phys. Rev. B* 10 (1974) 2039.
127. T. J. Smith, H. Capellmann, R. K. Kremer, K.-U. Neumann, K. R. A. Ziebeck, *Eur. Phys. J. B* 21 (2001) 341.
128. E. Paris, C. W. Nicholson, S. Johnston, Y. Tseng, M. Rumo, G. Coslovich, S. Zohar, M. F. Lin, V. N. Strocov, R. Saint-Martin, A. Revcolevschi, A. Kemper, W. Schlotter, G. L. Dakovski, C. Monney, T. Schmitt, *NPJ Quantum Mater.* 6 (2021) 51.
129. M. A. Green, M. Kurmoo, J. K. Stalick, P. Day, *J. Chem. Soc. Chem. Commun.* 17 (1994) 1995.
130. M. Hase, Y. Sasago, K. Uchinokura, G. Kido, T. Hamamoto, *J. Magn. Magn. Mater.* 140 (1995) 1691.
131. Y. Ando, J. Takeya, D. L. Sisson, S. G. Doettinger, I. Tanaka, R. S. Feigelson, A. Kapitulnik, *Phys. Rev. B* 58 (1998) R2913(R).
132. J. Takeya, I. Tsukada, Y. Ando, T. Masuda, K. Uchinokura, I. Tanaka, R. S. Feigelson, A. Kapitulnik, *Phys. Rev. B* 63 (2001) 214407.
133. B. Salce, L. Devoillea, R. Calemczuk, A. I. Buzdin, G. Dhalenne, A. Revcolevschi, *Phys. Lett. A* 245 (1998) 127.
134. Y. Chen, M. Sato, Y. Tang, Y. Shiomi, K. Oyanagi, T. Masuda, Y. Nambu, M. Fujita, E. Saitoh, *Nat. Commun.* 12 (2021) 5199.
135. M. Udagawa, H. Aoki, N. Ogita, O. Fujita, A. Sohma, A. Ogihara, J. Akimitsu, *J. Phys. Soc. Japan* 63 (1994) 4060.
136. W. Brenig, *Phys. Rev. B* 56 (1997) 2551.
137. R. N. Hall, *Phys. Rev.* 87 (1952) 387.
138. J. Ren, Q. Wu, H. Tang, G. Hong, W. Zhang, S. T. Lee, *J. Mater. Chem.* 1 (2013) 1821.
139. J. K. Feng, M. O. Lai, L. Lu, *Res. Bull.* 47 (2012) 1693.
140. W.-J. Meng, M. Zhao, H.-X. Yang, Y.-Q. Wu, H. Pu, R.-Z. Gao, Y. Yang, D.-L. Zhao, *Ceram. Int.* 46 (2020) 9249.

8 Conclusion and Future Direction

The book sheds light on transition metal oxides (TMOs) at low dimensions in Chapter 1. We discuss the surface properties of TMOs at lower dimensions, covering the effect of polarity, degree of coordinative unsaturation, and defect sites. Modification of electronic band structures at low dimensions is conferred for the expected changes in the magnetic, electrical, and optical properties. The electronic and magnetic properties of TMOs at lower dimensions are also elaborated, and the need for Mott-Hubbard approximation and the use of Heisenberg model were discussed. Modulation of bandgap and optical absorption in various TMOs at lower dimensions are also described briefly.

Chapter 2 deliberates on van der Waals (vdW)-bonded and engineered 2D TMOs for their structural evolution, optical and electrical properties, and possible applications. We discuss WO_3, MoO_3, and V_2O_5 as 2D vdW-bonded TMOs showing extraordinary optical and catalytic properties. A monolayer 2D TiO_2 is the only non-vdW-bonded TMO. 2D TiO_2 showed enhanced photocatalytic activity. Both non-vdW and vdW-bonded TMOs are still in an exploratory stage as growing these materials in large areas and on a large scale is the major challenge. Nevertheless, the ultrathin 2D TMO layers with noble optical and electrical properties have a significantly high possibility of device applications.

We discuss charge-density wave (CDW) in the context of quasi-1D TMOs in Chapter 3. Electron-phonon interactions leading to CDW are narrated for 1D metallic blue bronze ($K_{0.3}MoO_3$). The presence of the quasi-1D stripe of domains with finite width, involving dynamical modulation of spin and charge, is discussed for hole-doped superconducting cuprates and Ruddelsden–Popper-type layered structure of La_2NiO_4. The study of CDW originating from the antinodal region of cuprate Fermi-surface throws new light on the presence of pseudogap in high-temperature superconductivity. The continuous increase of checkerboard like periodicity of electronic modulation in $Bi_2Sr_2CaCu_2O_{8+\delta}$ with hole density supports a strong correlation of checkerboard appearance with the CDW formation. The pseudogap demonstrates various aspects of superconducting and normal states of a high-temperature superconductor. A complete understanding of CDW elucidating varieties of physical phenomena, however, is still in the early stage and needs further attention. Chapter 3 also covers various energy applications of electrochemical and photovoltaic energy

DOI: 10.1201/9781003623694-8

storage for TMOs like V_2O_5, bare and MnO_2-coated $Ni_xCo_{3-x}O_4$, hierarchical Co_3O_4/NiO nano-arrays, coated or core-shell Co_3O_4@MnO_2, Co_3O_4@NiO, ZnO@MnO_2, TiO_2, $NiCo_2O_4$ (NCO) 1D nanostructures, and catalysis using NCO, WO_3 nanofibers-C/Cu_2O nanowire array, *p*-Co_3O_4, and Pt-assisted TiO_2, CoO nanowires. Gas-sensing applications are also deliberated using various 1D TMO nanostructures in the resistive mode with the advantage of unidirectional electrical conduction in the 1D structures of iron oxide, CuO, vanadium oxide, Co_3O_4, NiO, TiO_2, MoO_3, and WO_3.

TMO nanostructures at 0D in the form of quantum dots (QDs) and nanoclusters are reported in Chapter 4. Photocatalytic and energy applications of various 0D TMOs are deliberated along with the chemiresistive gas sensing applications. TMO-based QDs are attractive because of the abundant availability of TMOs in the earth cast, inherent non-stoichiometry of oxygen in the compound promoting chemical activity, relatively high carrier lifetime, and stability. TiO_2, Cu_2O, CuO, MoO_3, NiO, and spinel ferrite ($NiFe_2O_4$, $MnFe_2O_4$) QDs and their heterojunctions and nanocomposites with graphitic (g)-C_3N_4 are considered for these studies. Optical confinement in QDs and nanoclusters is potentially useful for photocatalytic and energy applications. These 0D nanostructures with significantly high effective areas showed unsaturated chemical bonds at the surface, leading to the lowering of activation energy which is suitable for gas-sensing applications.

Properties and applications of heterostructures comprising layers of 2D transition metal dichalcogenides (TMDs), graphene, and MXenes (metal carbide, nitrides, and carbonitrides) on TMOs are discussed in Chapter 5 with an expected enhancement of functional properties than those for the individual constituents. Modulation doping of 2D-TMD/TMO heterostructure exhibiting charge carrier injection and carrier transport in the defect-free 2D layer showed great promise in high-quality nanoelectronic devices. A high value of photoreponsivity and advanced energy storage applications in 2D graphene/TMO heterostructures are also deliberated. Various applications of MXene/TMO heterostructures, namely, supercapacitive energy storage and novel capacitive deionization process used for the desalination process, are elaborated.

We discuss the growth and novel physical properties of low-dimensional perovskite oxides in Chapter 6. Nanowires and nanotubes are considered for the 1D perovskite oxides containing transition metals. The growth mechanisms of both nanowires and nanotubes of perovskite oxide are not fully understood, so it is almost impossible to propose a general growth-structure-property correlation. It is understood that research in the field of 1D perovskite nanostructures continues to be application-oriented and progressing rapidly, so a complete assessment of this subject will be daunting. We also study the growth and exotic physical properties of transition metal containing 2D perovskite oxides down to a monolayer with exotic physical properties compared with those shown by their bulk counterpart.

A few special TMOs, for example, VO_2 and $CuGeO_3$, showing Mott-Hubbard and spin-Peierls transition were discussed in Chapter 7. These materials are reported to possess 1D Heisenberg spin ½ chains leading to spin-Peierls type structural phase transition (SPT). A metal-insulator transition (MIT) and SPT are observed simultaneously in VO_2, where first-order Mott-type MIT is conjectured to initiate a second-order SPT without any observed latent energy. Spin wave (collective spinon vibrations) is recorded for both VO_2 and $CuGeO_3$. Pressure-induced phase transition in $CuGeO_3$, depending on the nature of the pressure medium, is also elaborated in this chapter. Along with the associated physical properties, applications of these materials are also explored.

TMO, being a functional material, plays an important part in the progression of modern civilization. Dimensionality is a knob to tailor its electronic, magnetic, and optical properties. It is definitely an opportunity where more work needs to be done, and insights gained may guide future efforts in these materials. The need to continue research in this promising direction is emphasized in connection with the unique possibilities of TMO nanostructured materials in the development of next-generation energy-harvesting and other futuristic device applications.

Index

A

Å, 13, 15, 35, 36, 38, 39
AAO, 91
Abarca, R. R. M., 30
Abazari, R., 27
Abbas, Q., 6
Abbaspoor, M., 69
Abdah, M. A. A. M., 69
Abdulla, W. Al, 31
Abe, Y., 50
Abello, L., 30
Abideen, Z. U., 52
Ablation, 104
Aboulouard, A., 5
Absorption, 4, 11, 15, 17, 23, 55, 56, 58, 59, 61, 72, 94, 108, 117
Abundance, 55, 62
Abundant, 43, 63, 80–82, 118
AC, 41
Acceleration, 58, 59
Acetone, 12, 15, 19, 44–46, 48–50, 65, 66
Acetylacetonate, 104
Acharya, H. N., 32
Acharya, A., 69
Acid, 23, 44, 66
Acoustic, 49
ACS, 25–27, 29–31, 33, 51, 52, 67–69, 83–89, 112
Acta, 5, 27–30, 53, 83, 85–88, 100, 112, 115
Activation, 15, 23, 45, 46, 55, 56, 63, 79, 95, 118
Active, 7, 13, 16, 18, 22, 41–43, 45, 49, 56, 58, 59, 61, 62, 72, 74, 80, 81, 95
Activities, 22, 23, 44, 49, 55–58, 61, 65, 74, 77–79, 81, 112, 117, 118
Actuation, 79
Actuators, 27–29, 31, 52–54, 68–70, 87, 112
Adams, D. M., 110, 116
Addison-Wesley, 50
Adhikari, K. K., 29
Adil, M., 5
Adler, D., 5
Adomavicius, R., 99
Adsorbates, 76
Adsorbent, 78
Adsorption, 55–46, 49, 57, 58, 61, 65, 66, 73, 78
Adv, 5, 25–28, 30–33, 51, 53, 68, 69, 83–89, 97–100, 112, 114, 115
Aerkoci, A., 5
Aerogels, 78
Aetukuri, N., 114
AFM, 3, 91, 93
Ag, 44, 45, 48, 59
Agarwal, V., 29
AgCl, 44
Agents, 1, 63
Agglomeration, 66, 73
Aggregation, 78, 79
Agriculture, 44
Aguilar-elguezabal, A., 32
Ahmad, I., 5
Ahmad, M., 29
Ahmadi, M., 5
Ahmadi, V., 5
Ahmed, A. S., 5
Ahmed, B., 89
Ahmed, F., 68
Ahmed, T., 25
Ahn, B. T., 31
Ahn, C. H., 84
Ahn, J.-H., 84
Ahn, S., 114
AIP, 69, 112
Air, 7, 48, 59, 66, 72, 73
Ai, Z., 32
Ajayan, P. M., 25
Ajitha, B. 26
Akande, A. A., 53
Akay, S., 32
Akbari, M. K., 27
Akimitsu, J., 116
Al-Abadleh, H., 5
Alamos, 50
Alamri, S. N., 20, 31
Albonetti, C., 99
Albuquerque, A. R., 115
Alcohols, 79
Al-doped, 72
Alemaida, I. M., 5
Alexe, M., 98, 100
Alfe, D., 85
Algadi, H., 30
Algarabel, P. A., 99
Alghamdi, A., 30
Alharti, F. A., 30
Aliannezhadi, M., 69
Aliev, R. A., 104, 113
Alignment, 79, 105
Alipour, L., 69
Alivisatos, A. P., 98
Aljaafari, A., 68

Alkali, 34
Alkaline, 1, 43, 79, 90
Alkoxide, 90
Allain, A., 83
Allen, P. B., 51
Allocation, 74
Allotropes, 7
All-oxide, 94
Alloys, 2, 5, 28–32, 53, 54, 67–69, 84, 98, 99, 113
All-perovskite, 94
Almeida, R. S., 53
Almohammed, S., 26
Alpay, S. P., 100
Alsalme, A., 30
Al-Shahry, M., 32
Alshareef, H. N., 89
Altomare, M., 31
Alu, A., 28
Alumina, 22, 91
Aluminium, 91
Amano, F., 10, 27
Ambade, S. B., 88
Ambient, 44, 61, 110
Ambiguity-free, 74
American, 24, 41, 57, 60, 73, 75, 82, 92, 94, 109
Amine, 46
Amirthapandian, S., 52, 53, 112, 113
Ammonia, 46, 59
Amorin, 115
Amorphous, 11, 13, 81, 110
Amplitude–voltage, 96
Amsterdam, 5, 6
Anaerobic, 55
Analysis, 10, 13, 22, 56, 66, 95
Analyte, 15, 46, 49, 50, 66, 67, 73
Anane, A., 100
Anasori, B., 83, 88
Anatase, 21–23, 25, 48, 56, 58, 59, 65
Anatomy, 38
Andersen, K. E., 84
Anderson, P. W., 3, 5
Ando, Y., 50, 111, 116
Andreu, T., 68
Anees, P., 30, 68
Ang, L. K., 83
Angelis, D. D., 85
Ángel-sanchez, K. Del, 32
Angew, 26, 31, 32, 87
Angular, 111
Animal, 55
Anion, 3
Anisimov, V. I., 112
Anisotropy, 13, 15, 37, 40, 93, 95, 108
Anjum, D., 89
Annadi, A., 84
Annealing, 9, 74
Annihilation, 74
Anode, 20, 25, 43, 78–81
Anodization, 43, 91
Anomalous, 34, 35, 76
Anomaly, 40
Anselmi-Tamburini, U., 54
Antibacterial, 1
Antibiotics, 18, 55, 58
Antibonding, 102
Antiferroelectric, 91
Antiferromagnet, 3, 34, 35, 79, 93, 108, 111
Antifogging, 107
Anti-hysteresis, 76
Anti-interference, 61
Anti-localization, 76
Antinodal, 36, 117
Antiparallel, 3
Antiphase, 16, 17
Antiquated, 48
Anti-reflecting, 107
Antisite, 43
Anti-solution, 106
Anti-Stokes, 106
Antonio, D., 99
Antonius, C., 32
Antonucci, P. L., 31
Anuradha, K. N., 97
Anwer, A. H., 67
Anwer, H., 68
Aoki, H., 116
Akoi, K., 32
Aono, M., 84
AOPs, 55
APL, 84
Application-oriented, 118
Approximation, 3, 34, 35, 58, 61, 64, 103, 117
Aquaculture, 55
Aqueous, 78
Ar, 10, 13, 64, 110
Arachchige, H. M. M. M., 83
Arash, A. 29
Arbiol, J., 68, 69
Architecture, 20, 21, 41, 42, 48
Area, 1, 4, 9, 11, 12, 18–21, 35, 38, 40–43, 45, 48, 55, 56, 65–67, 78, 79, 117, 118
Areal, 42, 59, 92
Arima, T., 6
Arimoto, Y., 100
Armand, M., 31
Arrays, 15, 20, 41–44, 48, 54, 91–93, 118
Arrows, 36, 92, 106
Arsat, R., 53
Article, 34
Artificially, 3

Arul, K. T., 31
Arun, T., 30
Asahi, R., 32
As-assembled, 41
Ascribed, 65
Asghar, H., 6
As-grown, 17
Ashok, C., 5
Ashrit, P., 6
Ashrit, P. V., 113
Asl, M. S., 68
Asokan, K., 31
Aspherical, 37
Assembly, 48
As-synthesized, 40
Asthma, 66
Asymmetric, 22, 63, 64, 73
Atalay, F. E., 99
Atalay, S., 99
Atkin, P., 25
Atmospheres, 22, 72
Atomic, 8, 10, 13, 35, 38, 39, 43, 46, 72, 74–76, 104
  atomic-layered, 16
  atomic-level, 23
  atomic-resolution, 35
  atomic-scale, 35, 71
Atoms, 7–9, 13, 15, 16, 18, 22, 25, 35, 37, 38, 67, 72, 101, 102, 108
Atrazine, 62
Atrei, A., 113
Attachment, 11
Attention, 1, 2, 7, 14, 23, 25, 44, 48, 55, 80–82, 108, 109, 117
Attributed, 40, 56, 61, 75, 77
Au, 23, 65, 77
  Au-decorated, 15
Average, 38, 56, 59, 91, 92
Averitt, R. D., 114
Avsar, D., 99
Awschalom, D. D., 83
Axe, J. D., 5, 50
Axes, 16, 37
Axis, 12, 16, 101, 103, 104
Ayala, A. P., 27
Azadmanjiri, J., 83
Azam, A., 68
Azharudeen, A. M., 29, 108, 115
Azman, N. H. N., 69
Azmat, S., 5
Azuma, M., 50

B

Babapoor, A., 68
Babar, F., 6
Babu, I. M., 87
Baćani, M., 5
Backbone, 46
Bacterial, 55
Baddour-Hadjean, R., 30
Bae, D. J., 85
Bae, S. -H., 114
Baesso, M. L., 27
Baeumer, C., 85
Bag, 35
Bagués, N., 5
Bahari, A., 27
Bahrevar, M. A., 98
Bahreyni, B., 54
Baik, J. M., 53
Baik, S., 97
Bai, J. M., 69, 97
BaIrO, 36, 37, 40
Bai, S., 29, 54, 70
Bai, W., 26
Bai, X., 27, 53
Bai, Y., 86
Balakrishnan, J., 84
Balberg, I., 114
Balcells, L., 100
Baldissera, G., 27
Balendhran, S., 29
Balicas, L., 83
Balke, B., 26
Balls, 102
Balogun, M. S., 40, 51
Bando, Y., 31, 51, 85, 112
Bands, 2, 3, 8, 11, 14, 18, 23, 31, 35, 37, 51, 56, 58, 73, 77, 83, 85, 95, 102, 103, 105, 117
  bandgap, 4, 7, 8, 11–15, 18, 20, 22, 23, 48, 49, 55, 56, 58, 72, 76, 79, 95, 103, 105, 108, 112, 117
  bandgap-modulated, 110
  bandwidth, 2
Banerjee, S. K., 26, 97, 99
Banys, J. 99
Bao, D., 31
Bao, J., 68
Bao, N. Z., 97
Bao, S. H., 85, 115
Bao, Q., 28
BaoHang, H., 86
Baraldi, A., 85
Baran, J. D., 84
Baratto, C., 53
Barbaros, O., 85
Barberi, R., 98
Bardi, U., 113
Barla, A., 84
Barriers, 49, 80

Bars, 96
Barsoum, M. W., 88
Barthelemy, A., 100
Basha, M. H., 32
Bashir, S., 68
Basic, 1
Baskoutas, S., 30
Basque, V., 113
Baste, Y. R., 115
Bast, W., 99
Basu, R., 16, 29, 69, 103–106, 108, 112–115
Bathula, C., 30
BaTiO, 72, 73, 77, 90, 95
Batista, V., 88
Batteries, 4, 6, 13, 14, 20, 40, 42, 48, 62, 63, 71, 78, 79, 81, 101, 109
Batzill, M., 84
Baughman, R. H., 86
BCN, 18
Beads, 49
Beam, 13, 15, 77, 92, 95, 108
Bedell, K. S., 50
Behaviour, 3, 15, 17, 18, 37, 71–76, 90, 95, 105, 111
Behnia, K., 83
Bell, A. T., 30
Belle, B. D., 29
Bellingeri, E., 99
Bellino, M. G., 99
Bending, 11, 16, 22
Benecha, E. M., 53
Bento, A. C., 27
Bera, S., 69, 112, 113, 115
Berlin, 5, 32
Bernussi, A. A., 113
Berzeliusone, J. J., 11
Besra, N., 69
Beydaghyan, G., 104, 113
BFO, 91, 93, 95, 96
Bhabha, 112
Bhaskaran, M., 29
Bhat, A. K., 114
Bhat, N., 29
Bhat, S. V., 97
Bhatia, S., 5
Bhosale, N. Y., 27, 31
Bian, K., 31, 51
Bias, 35, 38, 39, 44, 76, 106
Bibes, M., 100
Bidirectional, 108
Biermann, S., 103, 112
Biesheuvel, P. M., 83
BiFeO, 77, 91
Bignardi, L., 85
Bilayers, 16, 17, 80, 104
Bimetallic, 41, 90
Binary, 2
Binder, 20, 42
  binder-free, 63, 108
  binding, 22, 72
Binions, R., 113
Bio, 35, 48
  biocatalytic, 79
  bioelectron, 87
  biomarker, 12
  biomedical, 6
  biosens, 87
  biosensors, 108
  biotechnol, 68
Bipolar, 80
  bipolaronic, 3
Biscarini, F., 99
Bisconti, F., 27
Bisht, P., 15, 29
Bisphenol, 18
Biswas, A., 99
Biswas, S., 105, 114
BiTiO, 77
Black, 20, 71, 79, 112
Blackman, C. S., 113
Blanchard, R., 115
Bleaching, 112
Blinc, R., 99
Bloch-Wilson, 3
Blocks, 20, 93
Blood, 108
Blue, 18, 34, 56, 58, 78, 110, 112, 117
Blum, R. -P., 17, 30, 113
Blunt-edge, 79
BM, 58
Bo, X., 54
Bo, Z., 88
Boateng, E., 26
Bobrysheva, N. P., 113
Boca Raton, 68
Boeckl, J., 83
Bolivar, J., 51
Boltzmann's, 106
Bonaccorso, F., 26
Bond, 1, 2, 7, 8, 10, 11, 13, 15, 17, 22, 55, 56, 82, 102, 117, 118
Bondioli, F., 31
Book, 1, 46, 61, 117
Borchers, J. A., 113
Borhade, A. V., 110, 112, 115
Borisenko, V. E., 26
Borophene, 7
Bottom-gated, 77
Bottom-to-top, 8
Bottom-up, 8

Bound, 8, 34
boundaries, 44, 93, 95
Bovornratanaraks, T., 87
Bowen, M., 100
Boyer, M. C., 50
Boyjoo, Y., 88
Bozzo, B., 100
BPA, 18
Bragg, 36
Brammertz, G., 28
Branches, 42
Bräuninger, S., 110, 116
Bravo, Y., 69
Breakdown, 17, 18, 55
Breath, 66
Breathers, 106
Breedon, M., 53
Brenig, W., 116
Bridge, 10, 15, 42
Brivio, J., 83
Broeckx, J., 30
Broken, 2, 34
Bronze, 34, 117
Brook, 54
Brookite, 21, 22
Brown, 6
Brückner, W., 112
Brudvig, G. W., 27
Brunclíková, M., 27
BT, 106
BTO, 90–93, 95
Bu, 98
Budai, J. D., 112
Buffer, 95, 107
Building, 20
Bulaevskii, 111
Bulk, 2, 7, 8, 13, 22, 48, 56, 65, 71, 74, 90, 93, 95, 110, 118
bulk-resistance-control, 45
Bull, 5, 32, 33, 53, 86, 100, 113–116
Buller, S., 29
Bunches, 47
Burghard, M., 29, 53
Burkhardt, S., 27
Burris, J. L., 98
Burstein-Moss, 58
Butane, 65
Butanol, 50
Buthylamine, 19
Butler, S. Z., 25
Butterfly, 96
Buttrey, D. J., 5, 50
Buzdin, A. I., 116
By-products, 44, 56
Byun, J. -H., 52
Byun, Y. J., 113
Byung, H., 85

C

Cada, M., 27
Ca-doped, 45
CaFeO, 95
c-AFM, 76
Cage-like, 45
Cai, L., 29
Cai, S. 97
Cai, Q., 67
Cai, W., 69
Cai, X., 26, 30
Cai, Y., 28
Cai, Z. Y., 115
Calcination, 22, 66
Calculations, 3, 35, 62, 65, 103, 106
Calemczuk, R., 116
Cambridge, 5
Camjayi, A., 113
Camouflage, 108
Campanini, M., 53
Camposeo, A., 53
Candidate, 18
Cannilla, C., 31
Cantalini, C., 53, 54
Cao, C., 114
Cao, F., 29
Cao, G., 50, 51
Cao, H., 83
Cao, J., 33, 114, 115
Cao, L., 25
Cao, M., 86
Cao, P. J., 69
Cao, Q., 54
Cao, X. H., 67, 99
Cao, Y., 31
Capacitances, 20, 40–43, 59, 62–65, 74, 108
Capacitive, 71, 82, 118
Capacitors, 62, 65, 71, 72, 76, 78, 79, 94
Capacity, 14, 20, 22, 40–43, 59, 62–64, 78, 79, 81, 82, 92, 111, 112
Capasso, F., 115
Capellmann, H., 116
Carbaryl, 79
Carbides, 7, 71, 118
Carbofuran, 79
Carbon, 13, 18, 20, 22, 28, 40, 42, 43, 46, 63, 68, 79, 81, 82, 85, 86, 115
carbonaceous, 58, 63, 81
carbon-based, 63
carbon-coupled, 81
carbon-supported, 79

Carbonitrides, 71, 118
Carcinogenic, 55
Carmine, 56
Carmona, F., 112
Carretero-Genevrier, A., 99
Carriers, 14, 23, 40–42, 48, 55, 56, 58, 61, 65, 71, 72, 74–76, 93, 95, 107, 118
Carvalho, A., 83
Casal, B., 30
Casalongue, H. S., 87
Castellanos-Gomez, A., 28
Castro, A. H., 86
Castro, I. A. de, 28
Castro Neto, A. H., 83
Catalan, 100
Catalysis, 5, 18, 43, 58, 62, 118
Catalysts, 1, 4, 8, 12, 15, 23, 43, 44, 46, 56–59, 67, 77–79, 117
Catastrophic, 55
Cathodes, 4, 14, 15, 20, 25, 40, 78, 79, 81, 101
Cations, 2–4, 62, 80
Cavalleri, A., 114
Cavallini, M., 99
c-axis, 35, 37, 40
C–C, 77
C-coated, 14
CDI, 71, 82
C-doped, 12, 66
CdS, 23
CDW, 34–37, 40, 117
CdZnTe, 94
Ce, 25, 45
Cecconi, T., 113
Cell, 1, 4, 13, 15, 16, 19, 21, 25, 28, 33, 35, 37, 43, 44, 48, 66, 71, 77, 78, 93, 95, 101, 108, 110, 113, 114
Cen, C., 84
Cen, K., 88
Centimetre-long, 40
Centres, 9, 23, 34, 96, 101, 108
CeO, 20, 21
Cerdán-Pasarán, A., 29
Céspedes, O., 99
Cevc, P., 99
CF, 63
CH, 18, 19, 46, 47, 49, 65, 66, 77, 108
Chae, B. -G., 112, 113, 115
Cha, G., 25, 33
Chaghi, R., 84
Chai, N., 26
Chains, 13, 14, 16, 37, 65, 101, 103, 104, 106–108, 111, 118
Chai, Y., 83
Chaker, 25
Chakhalian, 83
Chakrabarti, K., 67
Chakra, C. S., 5
Chalcogenides, 7
Chalker, 88
Chamber, 46
Chamberland, 50
Chanda, K., 69
Chandra, A., 25
Chandra, S., 97, 113
Chandrasekar, A., 31
Chang, C. W., 87
Chang, H. L. M., 113
Chang, J., 84
Chang, J. L., 87
Chang, K. S., 98
Chang, S. H., 88, 114
Chang, T. C., 85
Chang, X., 87
Channels, 11, 14, 45, 46, 72, 75, 77, 91, 94, 107
Chan, N. Y., 85
Chao, D. 31, 85
Chao, Z., 68
Chapters, 4, 7, 34, 71, 97, 117–119
Character, 1
Characterization, 1, 9–11, 13–17, 21–23, 41, 55, 73, 74, 76, 105, 107, 111
Charge, 2, 4, 8, 11, 14, 25, 34–37, 40, 43, 45, 48, 55, 56, 58, 59, 62, 63, 65–67, 71, 72, 74–77, 91, 93, 94, 105, 112, 117, 118
  charge-density, 117
  charge-discharge, 43, 64
  charge-ordered, 36, 37, 93
Charging, 20, 62, 63
  charging-discharging, 42, 43
Charpentier, P. A., 113
Chatraphorn, S., 87
Chatterjee, K., 50
Chattopadhyay, K. K., 69
Chaudhari, P., 69
Chauhan, J., 5
Checkerboard, 35, 36, 38, 39, 117
  checkerboard-like, 35
Cheetham, A. K., 113
Chemical, 1, 7, 8, 10, 12, 13, 15, 16, 18, 21–25, 41, 43, 44, 46, 48, 49, 55–60, 63, 65, 72, 73, 75, 82, 91, 92, 103, 118
Chemiresistive, 44, 118
Chemisorption, 46, 66
Chemist, 11
Chemistry, 5, 12, 13, 110
Chemo-chromic, 4
Chemo-conductivity, 48
Chemosphere, 30, 32, 67, 68
ChemSusChem, 28
Chen, A., 26, 27, 70

Chen, B., 26, 113
Chen, C. H., 30, 68, 113
Chen, C. -L., 31
Chen, C. T., 112
Chen, D. D., 30, 67, 87
Chen, F., 97
Cheng, B., 31
Cheng, C. W., 28, 51, 114, 115
Cheng, H. -M., 26, 31, 51, 86, 87
Cheng, J., 53
Cheng, L., 6, 87
Chenglu, L., 113
Cheng, P., 54, 87
Cheng, Q., 31, 51
Cheng, S., 26
Cheng, W. -L., 87
Cheng, X., 29, 30, 31, 83
Cheng, Y., 100
Cheng, Z. P., 67
Chen, H. M., 28, 54, 85
Chen, J. -L., 31, 68, 84, 86–88, 99
Chen, J. Z., 33
Chen, K. -J., 51
Chen, L., 29, 54, 84, 86, 89, 99
Chen, N., 69
Chen, P. H., 33, 86
Chen, Q., 26, 32
Chen, S. J., 26, 29, 54, 86, 115
Chen, T., 69
Chen, W., 30, 83, 87, 98
Chen, X. D., 27, 52, 68, 83, 113
Chen, X. H., 113
Chen, X. S., 99
Chen, X. Y., 97
Chen, Y. W., 26, 29, 32, 84–86, 88, 98, 115, 116
Chen, Y. X., 84
Chen, Y. Y., 98
Chen, Z., 26, 30–32, 68, 114
Cheong, K. W., 6
Cheong, H., 98
Cheong, S. -W., 115
Chernova, N. A., 28
Cheung, K. M. C., 114
Chevrier, V. L., 31
Chhowalla, M. 114
Chi, X., 54
Chi, Y., 114
Chiarello, R. P., 113
Chicken, 104
Chimiresitors, 49
Chiral, 44
Chlorobenzene, 65
Cho, H. -J., 53
Choi, J., 53
Cho, J. H., 31, 51
Cho, J. W., 53
Choi, K. S., 27
Choi, M., 84, 115
Choi, S. J., 52, 53, 88
Choi, S. W., 52
Choi, S. Y., 84, 115
Choi, W. S., 83, 84
Choi, Y. C., 98
Choi, Y. M., 53
Chong, W., 114
Chou, H., 50
Chou, S. L., 88
Chou, W. -C., 31
Choudhary, N., 113
Chowde, C. G., 25
Chromism, 13
Chromogenics, 6, 11
Chu, 84, 87, 114, 115
Chung, H. J., 84
Chung, H. V., 53
Chung, J. S., 87
Chung, T. K., 84
Ci, S., 87
Ciccacci, F., 6
Circles, 35–38
Cladded, 43
Clarendon, 50
Class, 36, 105
Clauws, P., 30
Cleavages, 2
Clinic\clinical, 11, 108
Clockwise, 95
Cloth-supported, 40
Cluster, 103
CMR, 94
  CMR-based, 94
CNAs, 20
CNLs, 72
Co, 1, 4
Co-annealing, 61
Coath, J. A., 113
Coating, 13, 16, 40, 42, 43, 73, 104, 107, 118
Coaxial, 43
Cobalt, 43, 46
Cobden, D. H., 112
Cochran, R. E., 98
Cody, C. C., 27
Coefficients, 17, 25, 55, 96
Coehoorn, R., 33
Coercivity, 76, 92, 93
Coexistence, 11, 50, 62, 91
Coey, J. M. D., 99
Coincides, 103
Cole, B., 27
Colla, E., 97

Collapse, 105
Colloids, 27, 29, 59, 68
Colombo, 26
Coloration, 11, 15, 20, 21, 25
Colossal, 1
Colours, 4, 11, 92, 107, 110
 colour-changing, 25
 coloured, 11, 55
Combinations, 8, 71, 42, 65, 80, 102
Comini, E., 32, 53, 68, 69, 83
Commercial, 19, 61
Community, 103
Compactness, 109
Comparable, 2, 7, 15, 41–43, 48, 56, 59, 63–66, 74, 80, 93, 108, 110, 118
Comparison, 19, 35, 47, 60
Compatibility, 8, 48, 72, 82
Compensate, 45
Complex, 8, 44, 55
Component, 62
Composed, 18, 108
Composites, 18, 20, 40, 45, 46, 49, 59, 63, 66, 67, 78–80, 82, 112
 composite-based, 67
Composition, 1, 2, 46, 49, 80
Compounds, 7, 8, 18, 34, 44, 55, 62, 65, 71, 104, 110, 111, 118
Computational, 9
Concentrations, 12, 14, 19, 23, 34, 43, 47, 58, 60, 66, 93
Concepts, 71
Concern, 55
Concina, I., 69
Conclusion, 117, 119
Condensed, 52, 59, 71, 83, 97–100, 112, 114, 115
Conditions, 45, 61, 77, 82, 104
Conductance, 35, 38, 39, 74–77, 108
Conduction, 11, 23, 35, 43, 44, 58, 63, 65, 72, 77, 94, 95, 118
Conductivities, 1, 3, 9, 14, 15, 23, 37, 42, 46, 58, 59, 62–65, 76, 78, 80–82, 96, 108, 111
Conduits, 41
Configurations, 42, 43, 46, 48, 61, 63, 71, 73, 76, 77, 81, 94, 97
Confinement, 4, 8, 63, 95, 118
Conformal, 94
Conformations, 42
Cong, H. -P., 52
Cong, S., 28, 68
Conjectured, 118
Connection, 13, 72, 119
Connects, 15
Consequence, 34, 37
Consequential, 72
Conserved, 78
Considerably, 94
Consistent, 22, 35, 36
Constants, 4, 7, 34, 56, 62, 74, 93, 96, 111
Constituents, 71, 118
Contact, 46, 76, 77
Contaminants, 55
Context, 36, 117
Continuous, 36, 48, 117
Continuum, 111
Contour, 100
Contraction, 93, 110
Contrast, 45, 72
Contributes, 45, 62, 63, 66, 94
Contributions, 74
Control, 1, 3, 4, 13, 14, 22, 23, 77, 104, 107, 109
Conventional, 3, 42, 62, 72, 74, 93, 107, 109
Conversion, 6, 23, 43, 44, 51, 64, 78, 80, 105
Converted, 104
Converts, 15
CoO, 3, 44, 46, 79, 118
 CoO-graphene, 79
Cooled, 110
Cooperative, 72
Coordinated, 13
Coordinative, 2, 117
Copper, 35, 42, 45
Co-precipitation, 58, 66
Copyright, 1–120
Core, 3, 20, 42
 core-shell, 63, 78, 91, 118
Corners, 9, 15, 16, 22
 corner-sharing, 13
Correlation, 1–3, 34–36, 40, 50, 71, 101, 103–105, 111, 117, 118
Corresponds, 11, 13, 16, 22, 35, 92
Corrosion, 44, 48
Cortigiani, B., 113
Corundum, 44
Coslovich, G., 116
Cossari, P., 27
Cost, 7, 44, 61, 64, 81
 cost-effective, 43, 63, 81
 cost-effectiveness, 4, 20
Côtø, M. J., 32
Cotta, M. A., 52
Coulombic, 2, 20, 37, 63, 77, 81, 103, 108, 112
Counter, 21, 43
Counterclockwise, 95
Counterparts, 7, 41, 48, 118
Countries, 107
Coupling, 1, 3, 4, 21, 23, 24, 34, 37, 46, 59, 77, 91, 101, 108
Covalent, 2, 7, 8
Coy, J. M., 112
CPs, 65

Cr, 4, 104
Crabtree, R. H., 27
Crawford, M. K., 50
CRC, 68
Cr-doped, 105
Creative, 16, 17, 36, 37, 39, 64, 75, 96, 102, 106
Critical, 76, 104, 106, 111
Crocombette, J. -P., 113
Crossover, 23, 76
Cross-sectional, 96
Crucial, 1, 13, 14, 20, 103
Cruz-lόpez, A., 32
Crystal, 1–3, 8, 9, 14, 15, 20, 21, 25, 45, 66, 79, 91, 102, 111, 113
Crystallinity, 13, 51, 65, 75, 91–93, 95, 108, 109
Crystallography, 9, 49, 101, 103
Crystals, 8, 22, 78, 100, 111
CrystEngComm, 27, 30, 51, 52
CsI, 110
Cubic, 3, 9, 12, 44, 95
Cu-Cu, 111
Cucumber-like, 40
$CuGeO_3$, 108–112, 115, 118, 119
Cui, C. X., 68, 86
Cui, J., 32
Cui, S., 84, 86
Cui, X., 87
Cui, Y. -J., 25, 28, 114
Cui, Z., 67
Cunha, A. F. da, 28
CuO, 2, 34–37, 44–46, 58, 65, 66, 80, 108, 118
  CuO-graphene, 80
Cuprates, 2, 34–36, 108, 117
Cupric, 80
Curiale, J., 99
Curie, 93
Currents, 13, 35, 40–43, 59, 63, 64, 68, 77, 79, 105, 111, 112
Curves, 47, 63
CV, 63
CVD, 10, 103, 104
  CVD-grown, 103
  CVD-synthesized, 76
C-VOCQD, 40
Cyclability, 82
Cycling, 14, 19, 20, 25, 40–43, 56, 59, 62–65, 78, 79, 81, 108, 112

D

Daeneke, T., 25, 28
Daghrir, R., 32
Dai, H., 86, 87
Dai, J. Y., 26, 85, 98, 115
Dai, M., 51
Dai, R., 28
Dai, S., 26, 86
Dai, Z., 28, 33
Dakovski, G. L., 116
Dalavi, D. S., 20, 27, 31
Dalton, 51
Dames, C., 114, 115
Damascelli, A., 115
Dangling, 7
Danho, I., 85
Danish, M. S., 5
Dao, T. D., 53
Dark, 38, 92
Darmawi, S., 27
Das, A., 29, 52, 69, 115
Dasgupta, A., 30, 68, 113
Dash, S., 52, 53, 112
Das, K., 67, 97
Das, S., 68
Data, 38, 39, 42, 73
Datta, R. S., 97
Datta, S., 28
d'Ávila, M. A., 53
Davis, C., 50
Dawber, 85
d-b, 1
DC, 13, 37, 91
DCF, 18
Debate, 104, 107
Deb, S. K., 6, 28
Decay, 78
Decolourization, 55
Decomposition, 61, 78, 79, 90
Decoration, 15, 18, 49, 59, 65, 66, 73
Decrease, 1, 4, 12, 40, 46, 48, 58, 63, 65, 74, 93, 96, 104, 105, 108
Defect, 2, 13, 22, 23, 43, 45, 59, 66, 72, 76, 94, 111, 117
  defect-free, 118
Degenerate, 3, 58
Degradation, 1, 4, 18, 55, 56, 58, 61, 62, 72, 73, 80, 110, 112
Degrees, 2, 11, 71, 91, 117
Dehydrate, 66
Deionization, 71, 82, 118
Delaire, O., 114
Delaminating, 22
Delbari, S. A., 68
d-electrons, 1, 3
Delhaes, P., 112
Delocalization, 44, 107
Demic, S., 5
Deng, D., 69
Deng, H., 28
Deng, J., 52

Deng, L., 85
Deng, P., 53
Deng, Q., 67
Deng, S., 28, 32, 86
Deng, X. Y., 67, 69
Densities, 2, 3, 19, 20, 22, 25, 34–36, 40–44, 50, 59, 62–65, 74, 75, 78, 79, 81, 82, 95, 103, 105, 107, 111, 112, 117
Depletion, 66, 94
Deposition, 10, 13, 15, 22, 77, 91, 103, 104, 109
Depression, 55
der, 7, 8, 33, 46, 62, 71, 72, 81, 83, 115, 117
Dernier, P. D., 112
Desai, R., 31
Desalination, 71, 80–82, 89, 118
Desanto, M. P., 98
Deshmukh, M. M., 114
Deshmukh, P. R., 31
Design, 20, 42, 69, 97, 107
Desorption, 57
Despite, 20
Detavernier, C., 27
Detection, 4, 15, 46, 48–50, 59, 60, 65–67, 79, 108
Detectors, 8, 90
Devan, R. S., 27
Devices, 4, 6–8, 11, 13, 14, 17, 19, 23, 42, 59, 63, 64, 71–77, 80–82, 90, 93–95, 97, 101, 105–109, 117–119
Devi, S., 31
Devoillea, L., 116
DFT, 43, 95, 103
D'Haenens, J. P., 112
Dhalenne, G., 115, 116
Dhara, S., 29, 30, 52, 53, 68, 69, 112–115
Dhayal Raj, A., 29
Dhingra, S., 84
Dho, J., 113
Di, A., 31
Diabetes, 12
Diagnosis, 49, 66
Diagonal, 35
Diagrams, 8, 16, 36, 41, 82, 101
Diameter, 64, 66, 90–93, 109
Diamond-anvil, 110
Diankov, G., 87
Diaz, H. C., 84
Díaz, R., 68
Dichalcogenide, 63, 72
Dichalcogenides, 71, 90, 118
Diclofenac, 18
Dielectric, 7, 74–77, 90, 93
Differential, 35, 38, 108
Differentiate, 11
Diffraction, 35, 110
Diffusion, 4, 14, 20, 23, 25, 40, 45, 46, 49, 59, 63, 66, 67, 78, 81
Di, J., 54
Dillon, A. C., 28
Dimension, 1–8, 10, 12–14, 16, 18, 20, 22, 24–26, 28, 30, 32, 36, 38, 40, 42, 44, 46, 48, 50, 52, 54, 56, 58, 60, 62, 64, 66, 68, 70, 72, 74, 76, 78–80, 82, 84, 86, 88, 90, 92, 94, 96, 98, 100, 102–104, 106, 108, 110–112, 114, 116–119
Dimerization, 101, 103, 104, 106–108, 111
Dimers, 102, 104, 105, 108
Diminished, 111
Dimmer, 35
Ding, B., 53, 88
Ding, F., 85
Ding, G., 33
Ding, X., 100
Ding, Z., 28, 86
Diodes, 9
Dioxide, 46, 48
DiPietro, R. S., 99
Dipoles, 2, 74, 75
Directions, 8, 13, 16–18, 22, 35, 37–39, 42, 44, 62, 72–74, 76, 77, 92, 96, 105, 108, 112, 117, 119
Discharge, 42, 62–64
Discoloration, 14, 15
Discontinuous, 48, 103
Discovery, 7, 11, 108
Disease, 49
Diskus, M., 13, 28
Dislocating, 23
Dispersed, 20, 48, 61, 66, 72
Dissertations, 51, 54, 112
Dissimilar, 77
Dissociation, 46, 62
Dissolution, 82
Distinct, 21, 38, 71, 92
Distortions, 3, 9–11, 16, 22, 34, 66, 96, 101, 103, 105, 110
Distribution, 1–120
Dittmann, R., 84
Diversity, 7, 55, 97
Dix, N., 84
Djafari-Rouhani, B., 30
Dlamini, L. N., 26
DMFT, 103
DNA, 4, 80
Dobrosavljevic, V., 51
Doettinger, S. G., 116
Dohout, P. K., 98
Dolz, M. I., 99
Domains, 34, 35, 72, 77, 92, 93, 95, 117
Dominants, 3, 23, 48, 74, 111
Donates, 35, 67

Donato, A., 31
Dong, B., 54
Dong, C. -L., 31, 52, 54
Dong, F., 29, 51
Dong, G., 31, 100
Dong, H., 54, 88
Dong, J., 29
Dong, S., 28, 97
Dong, X., 86
Dong, Y. P., 115
Dong, Z. M., 67
Doniach, S., 103, 112
Donor, 14, 23
Doong, R. A., 30
Dopant, 23, 72
Dope, 11, 12, 14, 15, 21, 23, 25, 32, 34–36, 38, 39, 45, 58, 66, 71, 72, 77, 79, 103–105, 111, 117, 118
  doping-induced, 107
Dorhout, P. K., 98
Dörr, K., 85, 99
Dots, 40, 55, 78, 118
Dou, S. X., 31, 86
Doubles, 3, 37, 65, 101, 102
  double-exchange, 3
  double-stranded, 4
Downward, 46, 73, 92
Dozzi, M. V., 23, 32
Drain, 76
Dravid, V., 100
Drawn, 81, 82
Drese, R., 28
Drewes, J. E., 67
Drillon, M., 112
Drogui, P., 32
Droopad, R., 99
Drug, 95
Dry, 48
DSCs, 43
d-shell, 1, 71
Du, A., 87
Dual, 25
  dual-photoelectrode, 43
Duan, J. L., 27, 86
Duan, G., 69
Duan, R., 32
Duan, Y., 86
Dubal, D. P., 68
Dubcek, P., 32
Dubertret, B., 26
Dubey, P. K., 87
Dubonos, S. V., 25
Du, C. -H., 31
Due, 1, 2, 4, 7, 10–20, 22, 23, 25, 58, 61, 62, 65, 76, 93, 95, 103, 105–107, 110, 111
Du, F., 88
Du, J., 114
Dumcenco, D., 83
Duo, L., 6
Dupont, J., 33
Du, P. Y., 97
Durability, 79
Durmus, Z., 5
Dutta, S., 67
Dutt, M., 5
D'Urso, B., 84
Du, W., 53, 54, 67, 87
Du, X., 85, 87
Du, Y., 28
Duy, N. V., 53
Dye, 4, 18, 43, 55, 56, 58, 112
  dye-degradation, 18
  dye-sensitized, 43
Dynamics, 13, 22, 25, 34, 35, 101, 103, 110, 117
$DyScO_3$, 95

E

Eames, C., 84
Earth, 4, 23, 90, 118
EBE, 13
Ebina, Y., 26
Eby, R., 52
Economic, 56, 78
Edge, 9, 13, 16, 22, 56, 93, 108, 115
  edge-shared, 13, 22
EDLCs, 65
EDOT, 41
EES, 40–42
Efficiency, 4, 11–13, 15, 18, 20, 21, 23, 25, 43, 55, 56, 58, 61–63, 78, 79, 81, 105, 107, 108, 112
EFM, 92, 93
Efros, A. L., 32
Egg, 104
Einarsrud, M. A., 98
Eisaki, H., 50
Eisenbeiser, K., 99
El-Sherbiny, I. M., 5
Ekspong, G., 83
Elastic, 34
Electric, 1, 4, 7–11, 13–15, 17, 18, 23, 32, 37, 40, 48, 51, 58, 59, 61, 63, 65, 72, 74, 76, 78, 80–82, 92–94, 101, 105, 107, 108, 117, 118
Electroactive, 42, 79
Electroanalysis, 55, 87
Electrocatalysts, 43
Electrochemical, 6, 8, 12–14, 19–21, 25, 40–42, 49, 56, 58, 61–63, 77–82, 107–109, 117
Electrochromism, 4, 6, 9, 11, 13–15, 20, 21, 25, 46
Electroconductivity, 25

Electrode, 1, 6, 13, 20, 21, 25, 40–42, 46, 59, 62–65, 72, 74, 78, 81, 82, 108
Electrodeposition, 13, 22, 91
Electrodes, 13, 20, 21, 40–43, 59, 63, 74, 78–81, 94
Electrolysis, 79
Electrolyte, 12, 42, 43, 59, 62, 63, 107
Electromagnetic, 4, 78
Electromechanical, 49
Electron, 1–3, 7, 8, 11, 13–15, 18, 23–25, 27, 35–37, 44–46, 50, 55, 61, 62, 64, 65, 67, 71, 72, 77, 79–81, 83, 85, 90–94, 97, 101–103, 105, 106, 108, 114, 117, 119
  electron-devouring, 74
  electron-electron, 34, 101
  electron-hole, 1, 23
  electronics, 5, 7, 20, 71
  electron-phonon, 34, 101, 117
  electro-oxidation, 80
  electrophoretic, 91
  electrosorption, 71
  electrospinning, 12, 19, 20, 91, 92
  electrospun, 45, 46, 48, 49
  electrostatic, 93
Electronegativity, 4, 7
Element, 4, 7, 15, 23, 25, 44, 77, 78, 111
Elhadadi, B., 5
Elhamri, S., 83
Elm, M. T., 27
Elsevier, 5, 6, 14, 19, 21, 47
Emamjomeh, S. M., 53, 54
Embedded, 63
Emery, V. J., 50
Emission, 56, 58, 94
Encapsulation, 20, 59
Encircled, 38
Encyclopedia, 6
Energetically, 34
Energy, 1–4, 6, 8, 11, 14, 17, 19, 20, 22, 23, 26–33, 35, 36, 39–43, 51, 52, 55, 56, 58, 59, 62–64, 69, 71, 72, 78–82, 85, 86, 88, 102, 103, 105, 106, 108, 111, 113–115, 117, 118
  energy-dependent, 35
  energy-harvesting, 119
Engelmayer, J., 83
Enjalbert, R., 29
Environment, 6, 19, 44, 48, 49, 55, 62, 71, 72, 74, 76, 77, 82
  environment-friendly, 62, 81
Enzymeless, 79
Eom, C. -B., 83, 84
Eom, W., 88
Epifani, M., 68, 69
Epilayer, 74, 77
epi-$SrTiO_3$, 74
Epitaxy, 77, 80, 91, 95
Equivalent, 2
Er, 88
Erol, M., 5
Ertekin, E., 115
ES, 23
Esser, B., 5
Estrella, L. L., 5
Etching, 61
Ethanol, 44–46, 48, 49, 56, 60, 65–67, 72, 73, 110
  ethanol-assisted, 56
Ethylenedioxythiophene, 65
Eusebe, H., 99
Evaporation, 13, 15
Evolution, 43, 79, 95, 96, 105, 117
Exchange, 3, 65, 111
Excitations, 11, 37, 60, 73, 111
Excitonic, 72
Exfoliation, 7, 8, 16
  exfoliation-mediated, 22
Exotic, 2, 3, 90, 118
Expansion, 34, 81
Experimental, 13, 56, 95, 103, 106, 111
Explain, 2, 3, 11, 17, 18, 45, 105–107
Exponent, 104
Exposure, 4, 18, 46, 48, 55, 56, 65, 67, 73
Extraction, 11, 39, 72
Extraordinary, 1, 7, 40, 117
Extrinsic, 75
Eyert, V., 112
Eyewear, 11

F

Facet, 17, 22
Facile, 109, 111
Faglia, G., 68
Fairchild, S. B., 83
Fallahazad, B., 83
Fan, C. G., 115
Fang, B., 29
Fang, H., 100
Fang, J. Z., 67
Fang, M., 31
Fang, Q., 68
Fang, S., 86
Fang, T., 87
Fang, X., 51
Fang, Y., 28, 32, 100
Fang, Z. Q., 29, 67
Fan, H. J., 28, 31, 51, 83, 85, 98, 99
Fan, J. Y., 67
Fan, S., 115
Fan, W., 115
Fan, X. W., 26, 32, 51

Fan, Y., 114
Fan, Z. Y., 26, 31, 113
Farhoudian, S., 68
Far-infrared, 13
Farmer, B., 52
Farneth, W. E., 50
Farokhipoor, S., 85
Farooqi, S. A., 28
Farzi-kahkesh, S., 29
Fatima, Z., 6
Fattah, A., 29
Fauqué, B., 83
F-doped, 56
Fe, 1, 4, 12, 18, 78, 79, 96, 104
  Fe-doped, 43, 66
Fecher, G. H., 26
Feedback, 35, 39
Feigelson, R. S., 116
Fei, H., 26
Fei, Z., 112
f-Electron, 51
Felser, C., 26
Femtosecond, 105
Feng, C., 54
Feng, G., 8
Feng, J. K., 97, 109, 115, 116
Feng, L., 54
Feng, M., 98
Feng, S. M., 83
Feng, X., 85–87
Feng, Y., 52, 70, 88
FeO, 23
Ferain, E., 99
Fermi, 3, 36, 37, 117
Ferrari, A. M., 31
Ferrari, B., 69
Ferreira da Silva, A., 27
Ferrimagnetic, 105
Ferrite, 118
Ferro, 1, 93
Ferroelastic, 95
Ferroelectricity, 72–77, 91, 92, 94, 95, 98
Ferromagnetism, 36, 105
  ferromagnetic-paramagnetic, 91
Ferroni, M., 53
Fert, A., 100
FESEM, 41, 64
FET, 73, 74, 76, 77, 94, 107
Few-layered, 64
FIB, 92
Fibres, 22, 43, 53, 63
Fibrous, 43
Field, 1–4, 13, 15, 51, 59, 61, 72, 74–76, 83, 92–94, 102, 103, 105, 107, 108, 111, 118
  field-effect, 8
  field-induced, 74
  field-sensitive, 76
Filho, A. G. S., 11, 27
Filho, J. M., 27
Films, 4, 6, 10–15, 17–19, 22, 23, 25, 27–29, 32, 44, 46, 47, 54, 66, 72–76, 78, 92–97, 99, 100, 103, 104, 107, 108, 113
Filters, 90
Finazzi, M., 6
Finck, D., 83
Fingerprints, 110
Finite, 34, 104, 117
  finite-size, 4, 101, 104
Fiori, G., 26
Firsov, A. A., 25
First-order, 93, 101, 105, 118
First-rate, 20
Fisher, 98
Fits, 36, 111
Fitting, 96
Fivefold, 43
Flake, 64, 73
Flame, 15
Flat, 36, 74, 75
Flexibility, 8, 12, 20, 78
Flexoelectric, 77
Flores, 5
Flow, 64, 105
Flower, 78
  flower-like, 109
Fluctuation, 66, 110
Fluidic, 95
Flux, 106
FM, 93, 94
  FM-PM, 91, 93
Foam, 20, 42, 79
Focus, 7, 92, 49
Fogle, W., 34, 50
Foil, 20
Folkerts, T. J., 50
Fong, D. D., 83
Fontcuberta, J., 84
Food, 49, 87
Forbidden, 14
Forces, 13, 76, 77, 93
Forest, 42
Formaldehyde, 45, 48
Formamide, 16
Formic, 23, 44
Formula, 108
Fornasiero, P., 85
Fortrie, R., 30, 113
Fouling, 44
Fourcade, B., 113
Fourfold, 65

Fourier, 35, 38
  Fourier-transform, 39
Fragments, 104
Fraha, M. A., 6
Framework, 22, 59
France, 50
Francis, 1–120
Fransaer, J., 88
Free, 20, 23, 40, 56, 58
Freedom, 71, 91
Freeland, 83
Free-standing, 20, 40, 79, 96
Freire, P. T. C., 27
Freire, V. N., 27
Frequency, 11, 13, 16, 17, 22, 92, 93, 104–106
Freund, H. -J., 30, 113
Frozen, 111
Fruits, 79
FT, 35
Fu, D. J., 114, 115
Fuels, 4, 30, 43, 48, 78
Fu, H., 29, 32
Fujimori, A., 83
Fujishima, A., 32
Fujita, M., 116
Fukui, K., 28
Fulari, A. V., 69
Fularz, A., 26
Fu, L. H., 27
Functionalization, 1, 6, 7, 43, 45, 46, 48, 50, 56, 62, 65, 66, 71, 72, 77, 80, 81, 90, 93, 95, 97, 103, 118, 119
Fu, Q. S., 83, 113
Fursina, A., 85
Fu, X. L., 29, 99
Fu, Y. S., 53, 67
Fu, Z., 68

G

Ga, 104
Gabay, M., 83
Gadaleta, A., 99
Gaewdang, T., 26
Gaigneaux, E. M., 13, 28
Gaikwad, V. B., 115
Galceran, R., 100
Galeyeva, A., 6
Gallais, Y., 83
Galstyan, V., 83
Galy, F., 86
Galy, J., 29
Gan, L., 25
Ganduglia-Pirovano, M. V., 30
Ganesan, K., 68
Ganganagappa, N., 113
Gao, J., 88
Gao, L., 69
Gao, M., 67
Gao, P., 53, 86
Gao, R. -Z., 116
Gao, S., 29–31
Gao, W., 97
Gao, X. T., 86, 88
Gao, Y., 31, 51, 54, 87, 114
Gao, Z., 67
Gaps, 23, 34, 35, 37, 38, 42, 56, 103, 108
Garche, J., 6
Gargiani, P., 84
Gariglio, S., 83
Garnet-like, 110
Garoufalis, C. S., 30
Gartner, M., 28
Garzón, L., 100
Gases, 4, 6, 8, 12, 13, 15, 17–19, 22, 44–46, 48, 49, 54, 59, 65–67, 72, 79, 80, 101, 108, 110, 118
  gas-sensing, 15, 44–46, 49, 55, 65, 118
Gate, 72, 74–77, 94, 107
  gate-bias, 77
  gate-dependent, 77
Gaudon, M., 33
Gaussian, 36
Gautam, U. K., 85, 112
Gázquez, J., 84
Gazzadi, G. C., 99
G-band, 77
GC, 18
$GdScO_3$, 95
Ge, 53, 103
Geim, A. K., 25
Genç, A., 68
Gene, 4
Genenko, A., 83
Genevet, P., 115
GeO, 108
Geometrical, 111
Georges, A., 112
Gerlach, U., 112
Gesheva, K. A., 28
Ghadimi, L. S., 68
Ghahari, F., 84
Gharbi, N., 113
Ghasemi, S. A., 27
Ghatak, J., 52
Ghidiu, M., 88
Ghivelder, L., 99
Ghoreishi, F., 5
Ghosez, P., 83
Ghosh, B., 97
Ghosh, S., 69, 115

Giacometti, V., 83
Giancaterini, L., 53, 54
Giant, 95, 96
Gibbs, 2
Gigli, G., 27
Gilbert, D. A., 113
Gilje, S., 53
Gillaspie, D. T., 99
Gity, F., 84
GIXRD, 47
Glass, 3, 12, 20, 65, 104, 107
Glide, 49, 108
Glucose, 15, 79, 80, 108
Glycerol, 22
Glycol, 65
Gnanasekaran, L., 67
GO, 18, 58
Gogotsi, Y., 25, 83, 88
Gogova, D., 28
Golberg, D., 31, 51, 85, 112
Goldberger, J. E., 25
Gomes, U. U., 27
Gómez-Herrero, J., 26
Gong, G., 100
Gong, H., 51
Gong, Y., 83
Goni, A. R., 115
Gonzalez, Z., 69
Goodenough, J. B., 3, 5, 101–103, 112
Gopalakrishnan, R., 28
Gopal, N. O., 22, 32
Gosele, U., 98
Got, 80, 111
Gotic, M., 32
Gou, X., 86, 87
Governed, 107, 111
Govind, 87
Govindaraju, S., 30
Gowda, C. C., 25
Gowthaman, P., 5
GPa, 110
Grabowska, E., 32
Gradation, 62
Gradient, 15, 93
Graf, T. 114
Grain, 22, 44, 93
Grande, T., 98
Granqvist, C. G., 26, 27, 112
Graphene, 6, 7, 18, 20, 40, 58, 66–67, 71, 72, 74–80, 90, 97, 107, 118
  graphene-aerogel-supported, 79
  graphene-coated, 78
  graphene-Cr, 78
  graphene-encapsulated, 79
  graphene-wrapped, 80
Graphite, 20, 78
Graphitic, 18, 56, 118
Grassian, V. H., 5
Green, 6, 56, 62, 87, 110, 111, 116
Greentree, A. D., 25
Greven, M., 50
Grey, 75
Griffiths, 3
Grigorieva, I. V., 25
Grigoriev, Y. V., 98
Grigoropoulos, C. P., 51, 115
Grinding, 56
Groot, R. A., 33
Grossman, J. C., 115
Ground, 34, 37, 111
Group, 7, 13, 15, 16, 21, 22, 40, 56, 75, 80, 81, 95, 101, 108, 109
Grow, 10, 90, 91, 95
Grown, 11, 12, 22, 41–43, 48, 49, 56, 58, 59, 61–66, 77, 78, 80, 81, 91–93, 95
Growth, 8–10, 13, 15, 21, 22, 43, 44, 48, 90, 95, 97, 103, 109, 113, 118
  growth-structure-property, 118
Gruber, H., 28
Gruner, G., 50
Gruverman, A., 83–85
Guan, C., 86
Guang-Xin, N., 85
Guan, M., 27
Guan, R., 32
Guan, Y., 54
Gu, C. D., 51, 87
Guertin, R. P., 50, 51
Gu, H. S., 29, 97
Gui, K., 51, 86
Guinneton, F., 113
Guirleo, G., 113
Guisinger, N. P., 83
Gu, J. H., 6, 100
Gulich, T., 70
Gu, L. L., 51, 88
Gultekin, B., 5
Gu, M., 97
Gündogmu, P., 68
Gunnarson, O., 50
Güntner, A. T., 70
Guo, C., 25
Guo, D., 52, 88, 105, 114
Guo, F., 61, 68, 98
Guo, H., 5, 68, 86, 114, 115
Guo, J., 28, 29, 89
Guojia, F., 30
Guo, M., 27
Guo, Q. G., 87
Guo, R. Q., 87

Guo, S., 87
Guo, W., 22, 32, 97
Guo, X., 52, 54, 88, 99
Guo, Y., 28
Guo, Z., 114
Gu, P., 30
Gupta, A., 97, 100
Gupta, J. A., 25
Gupta, S. M., 32
Gupta, V., 5
Gurbatov, S., 69
Gu, Q., 31, 97, 99
Gutiérrez, H. R., 25
Gu, W., 32
Gu, Z. B., 97, 100
Guzmán, S. S., 29
Gwon, H., 31, 51
Gysel, R., 97

H

HAADF, 96
Haas, C., 33
Hadipour, A., 28
Hadi, R., 68
Hafnium, 72
Hagenmuller, P., 112
Haglund, R. F. Jr., 112
Hai, F. I., 67
Haines, J., 116
Hai, X., 31
Hai, Z., 27
Hakim, M., 53
Hakim, S. A., 19, 30
Half-filled, 102, 103
Hall, R. N., 74, 76, 105, 116
Hallum, V., 113
Halonen, N., 53
Halpert, J. E., 51
Hamamoto, T., 116
Ham, D. J., 28
Hamiltonian, 3
Hammerl, G., 84
Hammud, H. H., 68
Hanaguri, T., 50
Han, C., 51, 85
Han, D., 29
Handbook, 6
Hanfland, M., 116
Han, G. R., 97
Han, J. K., 51, 85, 86, 88, 98, 114
Han, K. J., 112
Han, L., 97, 100
Hanlon, T. J., 113
Han, N., 70
Han, S. D. 69, 86, 98
Han, S. H. 85
Han, T. H. 88
Han, W., 22, 53
Han, X., 28, 32, 52
Han, Y., 53, 67
Han, Z., 69
Hao, J., 85
Hao, J. H., 85
Hao, Y., 28, 68
Hao, Z., 52
Haq, M., 69
Harada, N., 69
Hariram, M., 6
Harlow, R. L., 50
Hartnoll, S. A., 114
Hasani, A., 27
Hase, M., 108, 111, 115, 116
Hashaikeh, R., 53
Hashimoto, K., 33
Hashizume, H., 26
Haspel, H., 27
Hassan, R. Y., 5
Hassini, A., 99
Hatano, T., 114
Haverkort, M. W., 112
Hayat, K., 99
Hazard, 29, 32, 33, 44, 52, 53
HBNI, 112
HCCN, 61
HCHO, 15, 44
He, 27–29, 31–33, 53, 54, 68, 83, 86, 88, 100, 114
Head-to-head, 95
Healthcare, 49
Heat, 34, 108, 111
  heat-activated, 105
Hebard, A. F., 99
Hee, 85
Heidelberg, 5
Height, 37, 49
Heiman, D., 99
Heinz, T. F., 25
Heisenberg, 4, 101, 103, 104, 106, 107, 111, 117, 118
Helium, 110
Hellberg, C. S., 84
Helmer, F., 28
Hemamalini, R., 67
Hematite, 44
Hemberger, J., 83
Henrich, V. E., 5
Heo, J., 84
Heo, Y. J., 6
Herklotz, A., 85
Hermosa, C., 26

Hernandez, B. A., 86
Hernandez, Y., 98
Hernández-Magallanes, J. A., 29
Hernandez-Sanchez, B. A., 98
Herranz, G., 84
Hesse, D., 98
Hetero-epitaxial, 104
Heterogeneous, 44, 65, 79
Heterojunction, 45, 46, 48, 49, 56, 59, 61, 62, 65, 67, 71, 73, 74, 118
heterojunction-based, 62
Heterostructures, 7, 8, 23, 42, 45, 48–50, 56, 58, 63, 64, 71–77, 79–83, 85, 87, 89, 95, 97, 118
heterostructure-based, 72
Hexagonal, 10–12, 45, 62
hexagonal-BN, 76
HF, 12
Hg, 94
Hierarchical, 21, 42, 43, 45, 48, 49, 64, 80, 118
Hietschold, M., 98
Hieu, V., 53
High-crystalline, 61
High-dielectric-constant, 72
High-efficiency, 43
High-energy, 81
High-frequency, 93
High-magnification, 47
High-performance, 7, 20, 23, 41, 78, 79, 81
High-purity, 103
High-quality, 71, 108, 118
High-resistance, 94
High-resolution, 38
High-sensitivity, 8
High-temperature, 1, 117
Hippalgaonkar, K., 114
Hirose, Y., 69
Hjelm, A., 26
h-MoO, 12
Hng, H. H., 30, 31, 69, 85, 86
Hoa, L. T., 87
Hoa, N. D., 53
Hoffman, J., 84
Hofman, K., 53
Ho, J. C., 31
Ho, O., 115
Holes, 14, 23, 24, 34, 35, 36, 55, 56, 62, 67, 104, 105, 117
hole-doped, 34, 35
hole-rich, 105
Hollen, S. M., 25
Hollow, 40, 45, 46, 66, 78
Holtz, M., 113
Homi, 112
Homo, 45
Homogeneity, 104
Homogenously, 66
Homojunction, 58, 66
homojunction-induced, 66
Hong, B. H., 53, 84
Hong, C. K., 27
Hong, J., 51, 87, 114
Hong, G., 116
Hong, S. H., 52
Hong, S. J., 51
Hong, S. K., 98
Hong, S. S., 25
Hong, V., 53
Hong, X., 84, 85
Hopping, 3, 23, 103
Horizontal, 35
Hormoz, S., 114
Horrocks, B. R., 26
Hoshyargar, F., 52
Hou, C., 32, 97
Hou, G., 28, 31
Hou, H., 67
Hou, L., 53, 88
Hou, T., 52
Hou, X., 29
Hou, Y., 68
Hou, Z., 87
Howald, C., 50
HRTEM, 57, 82
Hsieh, H. H., 112
Hsu, J. F., 84
Hsu, T. K., 87
Hua, H., 114
Huang, B., 53
Huang, C. K. 18, 30, 112, 113
Huang, G., 69
Huang, J. Q., 25, 68, 87, 88
Huang, M., 51, 84
Huang, Q., 50
Huang, R., 29
Huang, S., 52, 115
Huang, T., 86, 87
Huang, W., 29, 86
Huang, X. J., 31, 52, 55, 68, 84, 86, 87
Huang, X. H., 87
Huang, X. L., 31
Huang, Y. C., 31, 51, 67, 86–88, 100
Huang, Y. S., 83
Huang, Y. Y., 52
Huang, Z. H., 69, 88, 89
Hua, W., 88
Hua, Z., 69
Hubbard, 3, 103
Hubicka, Z., 27
Hübner, U., 67
Hu, C., 33, 114

Hucho, C., 30, 113
Huda, N. U., 6
Hudson, E. W., 50
Hu, G., 28, 85
Hughes, G., 84
Hug, H. J., 5
Hu, H., 67, 86
Hui, Y. Y., 85
Huh, J. -S., 69
Hu, J. Y., 28, 86
Hu, M., 69
Humidity, 15, 46, 66
Hunter, B. A., 26
Hunter, S., 112
Hunt, M. R., 6
Hu, P., 68
Huo, L. -H., 29, 30
Huo, T., 29
Huo, X., 67
Hu, R., 27
Hurdle, 23
Hur, S. H., 87
Husbandry, 55
Hussain, C. M., 6
Husson, E., 30
Hu, W. D., 84
Hu, X. J., 29, 32, 87, 97
Hu, X. Y., 67
Huyan, H., 97
Hu, Y. M., 29, 86, 97
Hu, Z. L., 86, 97, 100, 112
Hwang, H. Y., 84
Hwang, I., 33
Hwang, J. Y., 84
Hwang, T. H., 31, 51
Hwang, Y. J., 84
HWO, 62
Hybrids, 41, 42, 63, 67, 77–80, 82, 95
Hyde, L., 27
Hydrazine, 59, 60
Hydrocarbons, 77
Hydrochloric, 66
Hydrochloride, 61
Hydrogel, 78
Hydrogen, 4, 7, 19, 22, 43, 44, 46, 52, 59, 66, 86, 108
Hydrolysis, 55
Hydrophilicity, 8
Hydrophobicity, 8
Hydrostatic, 105, 110
Hydrothermal, 8, 10–12, 16, 19, 22, 40, 41, 48, 49, 58, 59, 61–64, 90, 91, 103, 104, 109–112
Hydroxyl, 55
Hydrolysis, 77
Hye Ri, K., 85
Hysteresis, 74–76, 78, 96
Hyun, A. J., 85

I

Iannaccone, G., 26
Ibarra, M. R., 99
Ibrahim, A. A., 30
Idris, N. H., 87
Iglesia, E., 30
Ihm, K., 115
Ikuta, H., 50
Ilyas, S., 5
Imada, M., 83
Imaging, 35, 41, 47, 57, 64, 73, 82, 92, 96, 108
Imawan, C., 28
Immersing, 22
Impellizzeri, G., 6
Impurities, 23, 46
 impurity-driven, 23
Imran, M., 52
Inagaki, M., 89
Inbanathan, S. S. R., 30
Incommensurable, 34
Incommensurate, 37, 111
Indicate, 36, 47, 58, 92
Indirect, 14, 48, 49, 77
Indium, 104
Induced, 1, 3, 14, 34, 77, 93, 95, 103, 105, 106
Industry, 44, 55
Infrared, 90, 101, 108, 113
Infused, 34
Ingler, W. B. Jr., 32
Inhibitory, 62, 80, 82
Injection, 71, 72, 74, 106, 118
Inorganic, 56, 108, 110
In-oxide, 77
In-plane, 8, 36
Insulating, 3, 34, 37, 42, 46, 93, 94, 101–103, 106, 107
Insulators, 2, 3, 93, 94, 103, 105
Integration, 41, 42, 58, 94, 97
Intensity, 16, 72, 106, 111
Interaction, 2, 3, 8, 34, 44, 71, 72, 77, 80, 81, 97, 117
Intercalated, 78
Interfaces, 29, 30, 44, 46, 49, 51, 52, 56, 61, 65–68, 71, 72, 74, 76, 77, 80, 81, 85–89, 100, 112
Interfacial, 44, 56, 71, 76
Intermediate, 9, 68
Interplane, 8
Interpolated, 59
Intervalence, 11
Interwoven, 20, 40
Intra-ionic, 3

Intrinsic, 12, 14, 74, 76, 79, 80
Inverse, 105
inverse-spinel, 44, 45
Ions, 1, 3–7, 11, 14, 20, 23, 25, 28, 34, 45, 54, 59, 61–63, 71, 81, 92, 94, 96, 101, 104, 107, 108
ion-beam, 104
ion-diffusion, 80
Iqbal, M. A., 26
Iqbal, M. J., 99
Iqbal, Y., 99
Ir, 77
Iron, 44, 45, 49, 79, 118
Irradiation, 23, 43, 56, 110, 112
Irreversible, 107
Irvin, P., 84
Ishiwara, H., 100
Islam, M. S., 84
Ismach, A. F., 25
Ismail, M., 5
Isolated, 4, 91
Isopropanol, 10
Isotherms, 57
Ithurria, S., 26
Itinerant, 2, 3
Ito, S., 33
Ivanda, M., 32
Ivanova, T., 13, 28
Ivanov, I., 112
Ivanovici, S., 87
Ivanov, M., 99
Iwasaki, M., 33
Iwasawa, Y., 28
Iwasa, Y., 114

J

Jabeen, N., 85
Jacob, 11
Jacob-Dolan, J. A., 11
Jaiswal, M., 84
Jana, A., 83
Jana, D., 67
Janaky, C., 27
Janek, J., 27
Jang, A. R., 85
Jang, D., 53
Jang, H. W., 27
Jang, J. H., 31, 68
Jang, J. S., 52, 53
Jan, T., 5
Jantunen, H., 53
Jariwala, D., 25, 68
Jaroniec, M., 67
Jasem, S., 5
Javed, M. S., 16, 25, 30
Jayababu, N., 29
Jayachandran, M., 28
Jayaraj, S. K., 30
Jayaraman, A., 115, 116
Jayavel, R., 30
Jekel, M., 67
Jenifer, A., 18, 30
Jensen, I. T., 29
Jeon, B.-H., 30
Jeong, H. M., 31, 51, 84, 85
Jeong, J., 114
Jeong, K. O., 98
Jeong, S.-Y., 67
Jeon, K., 52
Jeon, S., 51
Jha, R., 29
Jia, D. C., 28, 98
Jia, F., 32
Jia, J., 53
Jiang, C., 52, 67
Jiang, D., 25
Jiang, J., 51, 68
Jiang, K. 115
Jiang, L., 31, 32, 51, 54, 83, 99
Jiang, Q. 31
Jiang, S. 67
Jiang, T.-J., 52
Jiang, W., 26, 113
Jiang, X. P., 53, 97
Jiang, Y., 87, 88
Jiang, Z. X., 31, 88, 100
Jian, J., 30
Jian, W., 114
Jiao, A., 69
Jiao, Y., 25, 87
Jia, Q., 27
Jiao, Z., 87
Jia, W., 28
Jia, Y., 115
Ji, B., 27
Ji, D., 69
Jie, W., 85
Jie, X., 70
Ji, G., 69
Ji, H., 27, 29, 33, 52, 53, 85
Jin, C. G., 97
Jin, D., 28, 53
Jin, F.-M., 87
Jinglian, B., 114
Jing, Y.-L., 114
Jin, H., 25, 53, 87
Jinhua, L., 113
Jin, J., 32, 83
Jin, M., 32
Jin, P., 115

Jin, Q., 51
Jin, R., 5
Ji, X., 68
Ji, Y., 26
Ji, Z., 88
Jnawali, G., 84
Jo, B. H., 114
John, 6
Johnston, S., 116
Johnston-Halperin, E., 25
Jones, K., 27
Jones, S., 52
Jong Hyun, A., 85
Jöns, 11
Joo, O. S., 68
Joraid, A. A., 31
Joshi, U. A., 97
Jouaiti, A., 5
Joule, 105
Jovic, V., 114
Jugdersuren, B., 99
Juliao, J. F., 27
Junction, 72, 73, 80
Jung, H.-T., 51, 53
Jung, J., 52
Jung, M. G., 51
Jung, S. W., 98
Jung, Y. S., 52

K

Kaciulis, S., 69
Kadam, A. V., 27
Kahya, O., 84
Kakihana, M., 29
Kalantar-zadeh, K., 25, 26, 28, 53
Kalcheim, Y., 113
Kalderis, D., 32
Kalinin, S. V., 112
Kam, K. C., 113
Kamalam, M. B. R., 18, 30
Kanamori, J., 3, 5
Kanehira, M., 114
Kaneko, N., 50
Kaner, R. B., 25, 26, 53
Kang, F., 87, 89
Kang, H. W., 51, 52, 74, 84
Kang, J. K., 31, 51
Kang, K. T., 31, 51, 74, 84, 86, 87
Kang, K.-Y., 112, 113
Kang, L., 85, 86, 114
Kang, M., 30
Kang, S., 99
Kang, T. T., 84
Kang, X., 33
Kang, Y., 27
Kan, H., 68
Kanjanaboos, P., 87
Kankala, R. Kumar, 69
Kanno, R., 6
Kansal, S. K., 67
Kapitulnik, A., 50, 116
Kappertz, O., 28
Karakitsou, K. E., 23, 32
Karthiga, R., 29, 115
Karthik, Y. P. V., 26
Karuppiah, G., 30
Kasap, M., 32
Kashan, H. S., 29
Kasirga, T. S., 112
Kataphinan, W., 52
Katoch, A., 52
Kats, E., 27
Kats, M. A., 108, 115
Katsura, H., 114
Kaur, A., 67
Kaur, D., 113
Kaur, N., 83
Kaushal, A., 113
Kawasaki, M., 114
Kawasaki, S., 99
Kaya, N., 99
Kayan, B., 32
KBr, 110
Kemper, A., 116
Ke, Q., 86
Ke, Y., 114
Khalid, H., 6
Khalil, A., 53
Khan, A., 31, 114
Khanal, D. R., 115
Khan, A. U. H., 6
Khan, S. U. M., 32
Khan, M. I., 115
Khan, M. Z., 68
Khataee, A., 32
Kheel, H., 54
Kholkin, A. L., 99
Khomskii, D. I., 5, 112
Khoo, E., 51
Khoshaman, A. H., 54
Kido, G., 116
Kieffer, J. C., 114
Kihlborg, L., 28
Kile, B. M., 32
Kim, B.-J., 112, 115
Kim, B. S., 68
Kim, C.-H., 53
Kim, D. H., 27, 52
Kim, D. J., 84, 98

Kim, D. Y., 26, 30
Kim, E.-S., 29
Kim, G., 115
Kim, H. K., 51, 85, 86, 98, 113
Kim, H.-S., 31
Kim, H.-T., 112, 113, 115
Kim, H. W., 52
Kim, I., 27, 30
Kim, I.-D., 52, 53
Kim, J., 68, 86, 98
Kim, J.-H., 27, 52
Kim, J. J., 53
Kim, J. K., 27, 115
Kim, J. S., 68
Kim, K.-H., 67
Kim, K. S., 85
Kim, M. D., 26
Kim, M. H., 53
Kim, N.-H., 52, 53
Kim, N.-Y., 29
Kim, P., 84
Kim, S.-D., 84
Kim, S. H., 88
Kim, S. J., 53, 88
Kim, S. M., 84
Kim, S. S., 52, 83
Kim, S. W., 30, 84, 87, 98
Kim, S. Y., 27
Kim, T. W., 26, 85
Kim, Y. H., 85, 116
Kim, Y. K., 68
Kinetics, 23, 61, 81
Kinks-type, 2
Kiri, P., 113
Kirkendall, 78
Kis, A., 83
Kisch, H., 23, 32
Kittiwatanakul, S., 87
Kivelson, S. A., 50
Klanurak, I. 87
Klar, P. J., 27
Klimov, V. A., 104, 113
Kment, Š., 27
Kmentová, H., 27
KNbO, 95
Ko, C., 114, 115
Koh, H.-J., 33
KOH, 43, 59, 63
Kohler, M., 70
Kohli, S., 98
Kohsaka, Y., 50
Koide, Y., 51
Ko, J. W., 87
Ko, K.-E., 114
Kokubu, A., 33
Kolmakov, A., 112
Kondo, T., 50
Kondo effect, 93
Kong, C., 114
Kong, D., 31
Kong, X., 26
Kong, Y., 67
Konstantinovic, Z., 100
Konya, Z., 27, 53
Koo, H., 114
Koo, W. T., 52, 53
Kordas, K., 53
Kormanyos, A., 27
Korotcenkov, G., 6, 52
Korotin, M. A., 112
Ko, S. H., 51
Kostur, V. N., 51
Kottegoda, I. R. M., 87
Kou, L., 32
Kouser, S., 5
Ko, W. Y., 33
Kozlov, M., 86
Krasnok, A., 28
Krautz, E., 28
Kravchenko, I., 112
Kremer, R. K., 115, 116
Krishnaiah, P., 30
Krishna, N. G., 52
Kroemer, H., 83
Krotkus, A., 99
Krupanidhi, S. B., 98
Krysa, J., 27
Kšírová, P., 27
Kuang, Q., 32
Kuball, M., 97
Kuchmizhak, A. A., 69
Kugler, V. M., 100
Kukkola, J., 53
Kukovecz, A., 27
Kulandaivalu, S., 69
Kulinich, S. A., 69
Kulkarni, S. B., 31, 70
Kumar, A., 29
Kumar, C. B. P., 30
Kumar, K. Y., 30
Kumar, P. S., 29, 83, 87
Kumar, R., 53, 69, 87
Kumar, S., 30, 67
Kumar, Y., 29
Kumbhakar, P., 25
Kung, H., 5
Kuno, M., 25
Kurmoo, M., 116
Kurt, B. Z., 5
Kwon, H., 115

L

La, 34
LaAlO, 72, 76
Lacovig, P., 85
Ladder-like, 16
LaFeO, 90
Laguta, V. V., 99
Lahav, M., 33
Lahiri, B., 25
Lai, C. W., 98
Lai, M. O., 116
Lai, W., 88
Lai, X., 51
Lai, Z., 100
Lamas, D. G., 99
Lambrecht, W., 30
Lam, D. J., 113
Lamoureux, B., 114
Landström, A., 69
Laney, S. K. 114
Lange, C., 32
La–Ni–O, 35
Lan, T., 67
La–O, 35
Larentis, S., 83
Large, 2, 4, 8, 11–14, 23, 37, 40, 41, 45, 46, 65–67, 71, 74, 76–78, 80, 93, 102, 108, 117
  large-scale, 103
Larson, C., 53
Laser, 8, 15, 60, 104–106
  laser-ablated, 65
  laser-induced, 105
Latent, 105, 118
Lattice, 13, 34, 35, 37–39, 45, 46, 66, 80, 91, 96, 108, 110
  lattice-mismatch-induced, 80
Launois, H., 112
Lau, S. P., 85
Law, 23, 106
Layer, 2, 8, 12–16, 18, 20, 23, 25, 35, 37, 40, 43, 46, 58, 59, 62, 65, 66, 71–74, 76–78, 80, 81, 90, 93–95, 97, 104, 107, 108, 117, 118
  layer-by, 40
LBMO, 91
LCMO, 91–94
LDA, 103
LDOS, 35, 38, 39
Leakage, 4
Lebedev, O. I., 98
Lecoeur, P., 100
LED, 9
Lee, C., 54
Lee, B.-S., 52
Lee, D., 51
Lee, D.-D., 69
Lee, D.-H., 50
Lee, D.-S., 69
Lee, G. R., 52
Lee, H., 52, 83, 84, 85
Lee, J., 31, 67
Lee, J.-H., 67, 84
Lee, J. K., 98
Lee, J. W., 31, 51, 83, 84
Lee, J. S., 28, 85, 97
Lee, J. Y., 88
Lee, K. U., 31
Lee, P. S., 33, 51
Lee, S., 30, 52, 54, 84, 114
Lee, S. A., 84, 85
Lee, S. G., 6, 84
Lee, S. H., 6, 28
Lee, S.-H., 28
Lee, S. K., 26
Lee, S. M., 69
Lee, S.-M., 28
Lee, S., T., 116
Lee, S. W., 52, 67
Lee, Y., 84, 85
Lee, Y.-S., 31
Lee, Y.-H., 84
Lee, Y.-W., 112, 115
Leedham, T. J., 100
Left, 35, 39
Lei, B., 98
Leichweiss, T., 27
Lei, D., 86
Lei, F., 31
Lei, H., 31, 68
Lei, J., 51
Lei, M., 99
Leino, A.-R., 53
Leipzig, 112
Lei, X., 31
Lei, Z., 15, 29
Lei, X., 31
Lei, Z., 85, 86
Lemaitre, Y., 100
Leng, J., 12, 28
Leng, M., 22, 32
Leng, X., 32
Leonard, S., 116
Leonelli, C., 31
Lerttraikul, K., 87
Le, Q. Van 27, 68
Level, 2, 14, 15, 20, 65, 72, 77, 102
Leyva, A. G., 99
Levy, J., 84
Levy, P., 99
LFO, 90

Lhuillier, E., 26
Li, 14, 79, 81
Li+, 14, 78, 81
Lian, B., 84
Liang, C., 51
Liang, H., 26
Liang, J. G., 29, 32, 86, 89
Liang, L., 11, 27, 97
Liang, P., 85, 86
Liang, S. J., 68, 83
Liang, T. X., 97
Liang, X., 69
Liang, Y. Q., 67, 84, 86–88
Liao, T., 32
Liao, M., 51
Liao, X., 100
LIBs, 14, 15, 19–21, 25, 40, 42, 78–81, 109, 111
Li, B. X., 31, 27, 86
Lichtenstein, A. I., 112
Li, C. L., 26, 69, 86, 88, 98, 113
LiCoO, 4
Li, D., 29, 51–54, 68, 70
Li, E., 69
Li, F., 31, 51, 69, 86–88
Lifecycle, 82
Lifetime, 55, 56, 58, 61, 118
Li, G. A., 98
Ligand, 2
Li, H., 5, 25, 26, 29, 30, 51–54, 83, 89, 98
Li, H.-C., 84
Li, H. X., 51
Li, H.-Y., 54
Light, 4, 6, 8, 11, 12, 15, 18, 22, 23, 43, 44, 56, 58, 59, 61, 62, 65, 72, 73, 77, 105, 108, 110, 117
  light-emitting, 9
Li-ion, 62, 101
Li, J., 26, 52, 53, 67, 85, 88, 100, 113, 144
Li, K., 31, 69
Li, L.-N., 51–53, 68, 69, 86, 88, 97–99
Li, M., 26, 30, 85
Limit, 2, 12, 15, 18, 46, 49, 50, 59, 65, 66, 95
LiMO, 4
Li, M. S., 114
Li, M. Y., 98
Lim, S. Y., 31, 51
Lim, T. M., 30, 85
Lim, T. W., 115
Lim, Y.-S., 112, 115
Lin, C. F., 32, 33
Li, N., 30, 87
Linden, K. G., 67
Lindsay, R., 50
Line, 35–37, 39, 73, 75, 111
Linear, 4, 46, 59, 74, 108
Lin, H. J., 26, 69, 88, 112
Lin, J., 31, 85
Lin, K. J., 33
Lin, L., 30, 86, 114
Lin, M. F., 116
Lin, S. Y., 67, 100
Lin, T., 84
Lin, W., 67
Lin, X., 83
Lin, Y., 85, 87
Lin, Z., 84
Lipatov, A., 83–85
Li, P. C., 31, 54
Li, P. G., 98
Lipton-Duffin, J., 52
Li, Q., 31, 51, 86, 88
Liquan, F., 114
Liquid, 16, 107, 110
Li, R. Q., 67, 87, 113
Li-S, 81
Liscio, F., 26
Lisesivdin, S. B., 32
Lisin, A., 5
Li, S. S., 27, 52, 53, 86, 100
Li, T., 69, 83, 85, 100, 114
Lithiation, 81
Lithium, 6, 72, 79
  lithium-ion, 4
Lithography, 76, 92
Liu, A., 68
Liu, B.-T., 26, 30, 68, 86
Liu, C.-Y., 28, 29, 54, 67, 69, 85, 88, 113, 114
Liu, D. F., 69, 98
Liu, F., 27, 28, 83
Liu, G., 68
Liu, H., 51, 68, 83, 87, 88, 100
Liu, H.-J., 99
Liu, H.-K., 87
Liu, H. W., 97
Liu, H.-Y., 27
Liu, J., 6, 18, 25, 30–32, 51, 52, 67, 68, 85, 86, 88, 115
Liu, J.-H., 52
Liu, J. L., 68
Liu, J. P., 6, 42, 51
Liu, J. S., 99
Liu, J. Y., 67
Liu, K., 107, 114, 115
Liu, L. F., 52–54, 68, 89, 98, 114
Liu, M., 25, 83, 85, 97, 100
Liu, N., 25
Liu, P., 6, 31, 51, 87
Liu, Q., 26, 31, 54, 68, 85
Liu, R., 87
Liu, S., 13, 28, 29, 87
Liu, T., 53, 69, 86

Liu, W., 69, 70, 88, 98, 114
Liu, X., 25, 31, 69, 88
Liu, Y., 30, 69, 85, 88, 113
Liu, Y. C., 26
Liu, Y. H., 67
Liu, Y. T., 83, 88
Liu, Z. C., 26, 53, 67, 68, 83, 85, 86, 88, 97, 98
Liu, Z. G., 52, 97
Liu, Z.-H., 85, 86
Li, W.-J., 29, 52, 54, 86, 100, 114
Li, X. B., 27–29, 31, 52, 54, 67, 83, 85, 87, 113
Li, X. G., 97, 98
Li, X. M., 85
Li, Y., 22, 26, 29, 32, 52, 68, 69, 83, 86, 88, 100
Li, Y. F., 100
Li, Y. Q., 25
Li, Z., 28–30, 52–54, 67, 69, 85, 88
Li, Z.-F., 86
Li, Z.-J., 28, 70
Lizzit, D., 85
LMO, 92
Loading, 42, 45, 56, 58, 59
Local, 2–4, 10, 15, 23, 34, 35, 37, 44, 76, 103
Logan, B. E., 52
Lohar, G. M., 69
Lokhande, C. D., 31, 68
London, 6, 68
Lone pair, 67
Long-life, 78
Longo, E., 115
Long-range, 3, 16
Long-term, 19
Loop, 93, 96
Lopes, A. M. L., 99
Loredo, S. L., 29
Lorenz, T., 83
Lorenzo, J. E., 5, 50
Lorenzo, V., 99
Lou, X. J., 31
Lou, X. W., 85, 100
Low-bandgap, 59
Low-capacity, 40
Low-cost, 56, 81
Low-dimensional, 20, 34, 37, 40, 90, 103, 109, 118
Low-frequency, 16, 37, 105, 106
Low-temperature, 35, 91, 101
Low-voltage, 78
LPCMO, 91–93
LSCO, 91
LSMO, 91–94
LTO, 35, 37
LTT, 35–37
Lu, A., 86
Luan, V. H., 87
Lu, B., 87
Luca, G., 31
Lucazeau, G., 30
Lu, C.-L., 113
Lu, G., 27, 54, 87
Lu, H., 84, 85
Lu, J., 30, 97
Lukatskaya, M. R., 88
Lu, L., 31, 51, 87, 114, 116
Lu, M., 86
Luna, M. D. G. de, 30
Luo, H., 31, 51, 114
Luo, R., 29, 54, 70
Luo, S. D., 98
Luo, W., 87, 97
Luo, X., 28, 29
Luo, Y., 26, 51, 52, 69, 98, 100
Luo, Z., 98, 100
Lupien, C., 50
Lu, R., 85
Lu, T., 89
Lu, W., 85, 99, 100
Lu, X., 33, 52, 54, 88, 89
Lu, X. F., 97
Lu, X. L., 97
Lu, Y. M., 26, 68, 83, 87, 100
Lu, Z., 30, 31
Lv, N., 28
Lv, W., 87
Lv, Y., 28
Lv, Z., 29

M

Macdonald, T., J., 114
Machatine, A. G. J., 53
Macromol, 53, 97
Macroporous, 42
Macutkevic, J., 99
Madanagurusamy, 30
Hernández-Magallanes, J. A., 29
Magén, C. 99, 100
Maghemite, 44
Magnelli, 104
Magnetically, 1, 3, 4, 7, 8, 37, 75, 76, 79, 91, 93, 94, 101, 105, 108–111, 117, 119
Magnetism, 3, 40, 71, 90, 105
Magnetite, 45
Magnetization, 4, 93
Magnetoconductance, 76
Magnetoelastic, 108, 110
Magnetoresistance, 1, 93
Magnetotransport, 74, 75
Magnetron, 10, 18, 91, 104
Magnification, 38, 47, 64, 82

Magudapathy, P., 112, 114
Mahapatra, P. L., 25
Mahjoub, A. R., 27
Mahler, B., 26
Mahmood, A., 67
Mai, L., 51, 88
Mai, W., 30
Mai, Y. J., 85, 87
Maier, W. F., 32
Maiorano, V., 27
Maiti, S., 69
Mäklin, J., 53
Malarde, D., 114
Mali, S. S., 27
Mallouk, T. E., 26
Malmgreen, S., 27
Malviya, K. D., 25
Mandru, A. O., 5
Mane, A. A., 18, 30
Mane, R. S., 69
Mangalaraj, D., 29
Manganite, 91
Mani, G. K., 30
Manivannan, A., 99
Mannhart, J., 84
Mansoor, Q., 5
Many-body, 71
Mao, D., 51
Mao, J., 84
Mao, S. S., 27, 87
Mao, Y. B., 97–99
Mapping, 35, 41
Maps, 35, 38, 39, 95
Marcellis, A. De, 53, 54
Mardare, D., 32
Marezio, M., 112
Marín, L., 99
Marré, D., 99
Marsen, B., 27
Martin, W., 85
Martinez, B., 100
Martínez, J. I., 26
Mascarenhas, A., 28
Masuda, T., 116
Mathew, A., 29
Mathur, B. K., 32
Matrices, 58
Matrix, 8, 20, 23, 40
Matsumura, M., 32
Mattern, N., 112
Matzen, S., 100
Mayorga-Martinez, C. C., 5
Mazare, A., 33
Mazor, Y., 28
Mazumder, A., 29
MB, 56, 58, 61
MBE, 95
McCall, S., 50, 51
McCarron, E. M., 50
McCarthy, E. K., 84
McCarthy, J., 6
McClure, D. S., 6
McComb, D., 5
McGeough, C., 84
Mckenzie, B., 32
McKeown, D., 99
Mechanically, 4, 8, 9, 12, 80
Mechanism, 11, 44–46, 48, 58, 62, 65, 67, 73, 103, 112, 118
Mechanochemical, 56, 110
Medina, A. N., 27
Mehdi, M. H., 6
Meher, S. R., 6
Mehmood, U., 6
Mehta, B. R., 29, 87
Mehta, S. K., 67
Mehto, A., 5
Meissner, D., 32
Mei, Y., 85
Mekawy, M., 5
Melo, F. E. A., 27
Memaran, S., 83
Membrane, 91
Memory, 11, 15, 19, 76, 77, 90, 94, 95
MEMS, 49
Mendes, J., 27
Meng, B., 29
Meng, D., 30
Meng, F., 30
Meng, F.-Y., 29
Meng, F. Z., 31
Meng, J. 51
Meng, H., 29
Meng, K.-Y., 5
Meng, Q., 84
Meng, W.-J., 116
Meng, X. J., 31, 88
Meng, Y., 69
Merbouh, N., 54
Mercury, 78
Merenda, P., 112
Mesoporous, 42, 48, 66, 108
Mestres, N., 99
Metal, 1–8, 10, 12, 14, 16, 18, 20–26, 28, 30, 32, 34–72, 74, 76–78, 80, 82, 84, 86, 88, 90, 92–94, 96, 98, 100–118
  metal-based, 43
  metal-doped, 34
  metal-insulator, 105, 118
  metal-ligand, 1

metalorganic, 59
metal-oxide, 44
metal-support, 44, 77
metal-to-insulator, 2, 17, 34, 76, 91, 101, 104
Metallic, 1, 3, 17, 18, 23, 34, 37, 76, 101–108, 117
metallicity-induced, 17, 107
Metastable, 22, 44, 49
Methane, 18
Methanol, 56, 65, 66, 110
methanol-assisted, 56
Methyl, 18
Methylene, 18, 56, 78, 112
Metric, 72
Metronidazole, 58
Meuris, M., 28
Mezzi, A., 69
Mg-doped, 58, 111
MgO, 4, 91, 93, 95
Mhaisalkar, S., 86
Miao, L., 53
Miao, Y.-E., 68
Miao, Z., 88
Micro-actuators, 90
Microbial, 44
Micro-blocks, 104
Micro-boxes, 20
Micro-electronic, 1, 90
Micro-emulsion, 10
Micrometres, 90, 91
micrometre-long, 22
Micro-nano, 42
Microns, 91
Micro-objects, 108
Micro-oscillator, 93
Micropipette, 79
Microscopy, 35, 76, 77, 93, 94
Microspheres, 40, 66, 92
Microstructures, 42, 49, 66
Micro-thermometers, 108
Microwave, 90
microwave-assisted, 10, 110, 112
microwave-hydrothermal, 90
Mid-gap, 35
Migas, D. B., 26
Migration, 56, 61
Miklos, D. B., 67
Miller, E., 27
Millis, A. J., 83
Mineralization, 80
Ming, L. C., 110, 115, 116
Ming, N. B., 97
Miniaturized, 74
Minnam Reddy, V. R., 26
Minor, A. M., 115
Mintcheva, N., 69
Mintun, P., 83
Mi, R., 86
Mirabelli, G., 84
Mirrors, 11, 20
Mirzaeian, M., 6
Misewich, J., 99
Mishchenko, A., 83
Mishina, E. D., 98, 100
Mishra, S., 69
Mismatch, 80
Misra, S., 50
MIT, 2, 3, 17, 34, 40, 46, 76, 80, 91, 93, 94, 101, 103–108, 118
Mitchell, A., 26
Mitchel, W. C., 83
Mitra, G. B., 32
Mitrovic, I., 88
Mitsui, T., 32
Miura, N., 115
Mixture, 13, 110
Miyake, M., 100
Miyauchi, M., 33
Mjejri, I., 33
ML, 72
ML-MoS_2, 72
ML-TMDs, 72
Mlyuka, N., 112
Mn, 4, 41, 78, 80, 93
MnFe, 59, 118
MnO, 3, 4, 15, 40–42, 62, 63, 77, 78, 81, 82, 91–93, 118
MnOOM, 62
Mo, 1, 13, 49, 60
Mobile, 11, 94
mobility, 3, 72, 76
Modafferi, V., 19, 31
Mode, 10, 11, 13, 16, 17, 22, 37, 46, 105, 106, 118
Model, 3, 4, 105, 106, 111, 117
Modernization, 44, 119
Modulation, 11, 25, 34, 36, 49, 71, 72, 76, 77, 94, 105, 107, 117, 118
Modulators, 8
Modulus, 110
MOFs, 59
Mohamed, A. R., 32
Mohamed, S. H., 13, 28
Mohammadi, M. M., 68
Mohl, M., 53
Moholkar, A. V., 30
Moiseev, N., 5
Mokoena, T. P., 53
Moldenhauer, W., 112
Molecules, 28, 46, 60, 66, 78, 94
Molinari, M., 84
Molis, G., 99

Molten, 90
Molybdenum, 49, 54, 72
Moment, 2, 4
Momentum, 111
Monaghan, 84
Monk, P., 6
Monney, C., 116
Monoclinic, 9–11, 21, 22, 40, 59, 63, 101, 102, 106, 110
Monolayer, 7, 8, 10, 72, 75, 79, 90, 95, 107, 117, 118
Monolithic, 78
Monovalent, 62
Montini, T., 85
MoO, 4, 7–9, 11–15, 34, 44, 49, 58, 59, 62, 72, 82, 117, 118
Moodenbaugh, 50
Moon, 27, 83
Morante, J. R., 68
Morellón, L., 99
Morikawa, T., 32
Morin, F. J., 112
Morozov, S. V., 25
Morphology, 1, 13, 18, 19, 22, 25, 45, 46, 49, 78, 79, 103
  morphology-controlled, 15
Morrison, F. D., 98, 100
Mortimer, R., 6
MoS_2, 20, 64, 72, 73
MoSe, 72
MOSFET, 107
Moshkalev, S. A., 87
Moskovits, M., 53
Mosuang, T., 53
Motaung, D. E., 53
MoTe, 72
Mott, 3, 5, 23, 32, 103, 105, 107, 112
Motta, N., 29, 52
  Mott-Hubbard, 34, 103, 117, 118
  Mott-type, 105, 118
Moumen, A., 69
Mounasamy, V., 18, 30
MoX_2, 72, 73
Moyer, 50
MR, 93, 94
MTFET, 107
MTJ, 94
Mu, R., 26
Muhire, E., 67
Mukherjee, M., 25
Müllen, K., 86, 87
Muller, D. A., 84
Multicomponent, 63
Multi-dimensional, 71
Multi-ferocity, 71
Multiferroic, 1, 90, 91, 95, 107
Multifunctional, 76, 97
Multi-graphitic, 20
Multilayer, 66, 80, 107
Multi-valency, 13, 18
Murali, N., 53, 112
Murali, S., 86
Muralidharan, G., 87
Murin, I. V., 113
Music, 32
Mutual, 63
Mwakikunga, B. W., 53
Mxene, 71, 80–82, 118

N

NaCl, 110
Nadimicherla, R., 54
NaF, 12
Nagao, N., 53
Nagaosa, N., 114
Nagarajan, V., 98, 100
Nagaraju, G., 33
Nag, J., 103, 112
Nagle, R., 84
Naguib, M. 88
Naik, R., 52, 83
Najafi-Ashtiani, H., 27
Nakabayashi, S., 100
Nakagawa, N., 84
Nakajima, A., 33
Nakamum, Y., 50
Nakamura, J.-I., 69
Nakamura, Y., 5, 50
Nakano, M., 107, 114
Nakazawa, H., 26
Nam, K. H., 51
Nambu, Y., 116
Namini, S., 68
Nano, 5, 6, 20, 25, 26, 29–31, 33, 51–53, 56, 67, 69, 83–87, 89, 95, 97–100, 112, 114, 115
Nano-arrays, 20, 42, 118
Nanobelts, 14, 15, 18, 20, 105, 109
Nanobridges, 92
Nano-building, 20
Nanocable, 20
Nanochanels, 92, 93
Nanoclusters, 46, 61, 65, 79, 118
Nanocomposites, 20, 49, 56–58, 66, 108, 110, 118
Nanoconstriction, 92
Nanocrystalline, 18
Nanocrystallite, 66
Nanocrystals, 66, 73, 110
Nanocube, 79, 80
Nano-dimension, 1, 22

Nanoelectronic, 71, 118
Nanofibers, 12, 45, 46, 48–50, 92, 118
nanofiber-based, 44
Nanoflakes, 22
Nanoflower, 18
Nanoforest, 42, 43
Nanogenerator, 48, 93
Nanographene, 59
Nanomaterials, 6, 20, 26, 27, 67, 78
Nanoparticles, 23, 45, 49, 59, 61, 62, 65–67, 72, 73, 77, 79, 81, 104
nanoparticle-assisted, 62
nanoparticle-based, 65, 66
Nanopatterned, 45
Nanoplates, 19
Nanoporous, 12, 91
Nanoribbons, 78
Nanorods, 18, 44–49, 93, 105, 109
Nanoscale, 2, 26, 30–32, 42, 44, 53, 83, 85, 88, 92, 97, 109, 112
Nanoscience, 4, 68
Nanoscopy, 68
Nanosheets, 4, 10–16, 18–20, 22, 43, 58, 63, 65, 78, 79, 81, 95
Nanospectroscopy, 68
Nanospheres, 64–66, 78, 92
Nanostructures, 1, 4–6, 17–21, 23, 40, 42, 44–46, 48, 51, 56, 63, 65, 80, 103, 104, 108, 109, 118, 119
Nanotechnology, 52, 53, 67, 84
Nanotree, 42
Nanotubes, 20, 43, 45, 49, 50, 73, 90, 91, 93–95, 118
Nano-urchin, 18
Nanowire-array-integrated, 42
Nanowire-graphene, 80
Nanowires, 22, 40–46, 48, 81, 90–93, 106, 109, 111, 118
Nan, T., 100
Naocrystalline, 79
Narrow-bandwidth, 3
Nasilowski, M., 26
Nataraj, D., 29
Navale, S. T., 69
Navale, Y. H., 70
$NaVO_2$, 20
$NaV2O_5$, 20
Navone, C., 30
Nayak, A. K., 32
Nayak, B. B., 6
Nazari, M., 5, 113
Nazarov, D. V., 104, 113
Nb-doped, 74
NCO, 41, 43, 118
N-doped, 63, 78, 79
Near-surface, 5
Nedaei, M., 68
Neel, 3
Negative, 35, 59, 63, 72–74, 108
Negra, M. D., 113
Neppolian, B., 30
Neri, G., 31
Nesper, R., 28
Nesterov, O., 100
Netherlands, 6
Nethravathi, C., 85, 112
Net-like, 48
Neumaier, D., 26
Neumann, K.-U., 116
Neuromorphic, 74
Neutron, 35, 36
Ngamcharussrivichai, C., 6
Ngaruiya, J. M., 28
Nga, T. T. T., 31
Ng, C. F., 31
Nguyen, T. P., 27
NH_2, 18
NH_3, 12, 15, 19, 44, 45, 48, 50, 59, 65–67, 79
NH_4VO_3, 104
Ni, 1, 4, 68, 79, 108, 113, 118
Ni-based, 59
NIBs, 20
Nicholson, C. W., 116
Nickel, 6, 42, 43, 48
NiCo, 41, 43, 118
Nicolosi, V., 88
Niebieskikwiat, D., 99
Niehus, H., 30, 113
Nielsen, B., 85
Nie, R., 87
Nie, T., 88
Nie, Y. F., 97, 100
NiFe, 59, 118
Niklasson, G. A., 27, 112
Nilsen, O., 28
Ni-M, 59
Ni-Mn-Co-O, 42
Ni-MOF, 59
Ning, T. Y., 54, 98
Ning, W., 87
Ningyi, Y., 113
NiO, 3, 4, 35, 42, 44, 48, 49, 58, 62–64, 79, 80, 117, 118
NiO–graphene, 79
Niobite, 72
Nishio, S., 29
Nitride, 7, 71, 80, 118
Nitrogen-doped, 18
Niu, B., 25, 33
Niu, C., 40, 51

Niu, J., 99
Niu, J. J., 97
Niu, W., 26
n-n, 66
NO_2, 12, 15, 18, 19, 44, 45, 48–50, 59, 61, 62, 66, 67, 73, 80, 103, 112
Nobel, 83
Noble, 23, 43–46, 65, 117
Noda, M., 99
Noguesa, J., 5
Noheda, B., 100
Noh, J., 68
Noh, S. H., 88
Noh, T. W. 98
Nonbonding, 4
Non-collinear, 3
Non-dispersive, 36
Non-equivalent, 15
Non-interacting, 103
Non-linear, 37, 90
Non-magnetic, 3, 101
Non-polar, 2
Non-stoichiometry, 55, 66, 107, 118
Nontoxicity, 11, 56, 78
Non-vdW-bonded, 8, 117
Non-volatile, 77, 90, 92, 94
Nori, S. K., 53
Nourmohammadi, A., 98
Novák, P., 28
Novan, 68
Novoselov, K. S., 25, 83
Nowotny, H., 115
Nowrouzi, R., 27
NPs, 44
n-SnO, 73
n-TiO, 65
n-type, 8, 14, 15, 23, 48, 49, 80, 93
Numerical, 34
Nurunnabi, M., 6
n-WO, 66, 67
Nygren, M., 112

O

Obara, H., 115
Obermeier, E., 28
Obradors, X., 99
Obrovac, M. N., 31
Octadecahedron, 56
Octahedral, 9, 13, 14, 22, 45
Octahedron, 101
Octylamine, 59
Odedairo, T., 87
OER, 43, 79
Ogale, S. B., 99
Ogihara, A., 116
Ogita, N., 116
Ogletree, D. F., 114, 115
Ohno, T., 32
Ohsaka, T., 32
Ohtani, B., 27
Ohtomo, A., 84
Ohwaki, T., 32
Oil, 42
Okuyama, D., 114
Okuyama, M., 99
Olejnícek, J., 10, 27
Oliveros, A. N., 18, 30
O'Mullane, A. P. J., 52
One-dimensionality, 103
Ong, C. K., 99
Onn, Z. N., 33
Ono, H., 69
Ono, K., 26
Ono, S., 50, 114
O–O, 10
Oppermann, H., 112
Optical, 1, 4, 7, 8, 10, 11, 13, 15–17, 20, 22, 23, 25, 31, 34, 37, 49, 58, 72, 77, 78, 90, 92–94, 101, 107–109, 117–119
Optoelectronic, 8, 11, 13, 72
Orange, 56
Orbitals, 2, 4, 7, 11, 23, 71, 77, 91, 103
Ordering, 2, 3, 15, 16, 23, 36, 37, 48, 49, 61, 65, 66, 91, 93, 95, 103, 104, 105, 107, 111
Organic, 4, 12, 18, 40, 43, 44, 55, 65, 79, 109
Orhan, K., 85
Orientation, 43, 74
  orientation-dependent, 104
Oró, J., 99
ORR, 78, 79
Orthogonal, 12, 35, 106
Orthorhombic, 9, 13–16, 49, 91, 93, 95, 110, 112
Osada, M., 25, 26
Oscillation, 16
Osmolovskaya, O. M., 113
Ostrikov, K., 88
O-Ti-O, 22
Ott, H.-R., 51
Ou, J. Z., 26, 28
Ouma, C. N. M., 53
Ou, Q., 28, 29
Ouyang, C., 51
Ouyang, L., 97
Ouyang, Q., 86
Ovchinnikov, D., 83
Overlapped\ing, 2, 103
Overpotential, 43
O–W–O, 11
Oxford, 50

Oxides, 1, 2, 4–8, 10, 12, 14, 16–18, 20–22, 24, 26, 28, 30, 32, 34–72, 74, 76, 78–80, 82, 84, 86, 88, 90–118
  oxidation, 1, 4, 15, 18, 43, 44, 46, 55, 58, 59, 61–63, 78–80
  oxide-based, 44, 95, 107
Oxygen, 2–4, 7, 10, 12–15, 17, 22, 37, 43, 44, 55, 66, 74, 78–80, 101, 104, 105, 108, 118
Oxytetracycline, 58
Oyanagi, K., 116
Ozkan, S., 33
Öztürk, A., 68
Özyilmaz, B., 84

P

Pacchioni, G., 27
Pachoud, A., 84
Padture, 98, 99
Pakornchote, T., 87
Palacios, T., 26
Palange, E. 53, 54
Pale-green, 110
Pallecchi, I., 99
Palmisano, L., 31
PAM, 91, 93
Pan, A., 31
Panagopoulos, C., 83
Pan, B., 27
Pan, C., 28
Panda, D., 52
Pandey, R., 5
Pandian, R., 114
Pandya, S., 85
Pan, F., 86
PAni, 65, 67
  PAni-WO, 67
Pan, J. H., 31, 51, 70
Pan, K., 32
Pan, L., 5, 88, 89
Pan, M., 113
Pang, Z., 87
Pan, X., 97, 100
Pan, Y., 83
Panzera, G., 31
Papakonstantinou, I., 114
Parale, V. G., 69
Parallel, 3, 101
Paramagnetic, 105
Parashuram, L., 18, 30
Parida, S., 52
Paris, 116
Parisi, F., 99
Park, C. B., 85, 87, 114
Park, H., 97, 114
Park, H. H., 69
Park, H. K., 97, 99
Park, H.-S., 51
Park, H. Y., 30
Parkin, I. P., 113, 114
Parkin, S. S. P., 114
Park, J., 68, 83, 84, 114
Park, J. H., 112
Park, J.-W., 51, 67
Park, J. Y., 52
Park, N., 84
Park, S., 26, 54
Park, S. E., 54
Park, S.-J., 6
Park, T. J., 98
Park, Y. C., 31
Partial, 17, 37, 55, 73
Particulate, 56
Parvez, K., 86
Passive, 21
Pastoriza, H., 99
Patel, M., 68
Patil, P. S., 27, 31
Patil, R. A., 27
Patil, R. S., 27
Patil, U. M., 31
Patil, V. B., 69, 70
Patsha, A., 52, 113
Paudel, T. R., 83, 97
Paul, T., 69
Paula, N. H. de, 115
Pauli, 105
Pawar, S., 68
Pazhanivel, T., 29
Pb, 35
PbTiO, 76, 90
PCMO, 91, 93
p-CuO, 65
Pd, 46, 50, 65, 66, 94
  Pd-functionalized, 48
P-doped, 58
Pd-WO, 66
pdπ, 17
Pedersen, T. P. L., 28
PEDOT, 20, 40, 41, 65
Pegg, I. L., 99
Peierls, 34, 101, 105
Peierls, R. E., 50
Peierls-type, 34, 101
Pei, K., 26
Pei, L. Z., 109, 115
Pei, Y. Q., 115
Pellacani, G. C., 31
Pellegrin, E., 84
Pellegrino, L., 99

Pemble, M. E., 113
Peña, L., 100
Peng, B., 100
Peng, H., 32, 52, 86
Peng, J., 31, 51, 88
Peng, L., 86
Peng, S.-L., 70
Peng, X., 53, 86
Peng, Z., 26
Pennycook, S. J., 100
Pentoxide, 46
Perfecto, T. M., 10, 27
Pergament, A., 114
Periodically, 1, 34, 72, 101
Periodicity, 3, 36, 117
Perlstem, J. H., 50
Permeability, 9, 11
Perovskite, 1, 2, 4, 7, 12, 35, 71, 90, 91, 93, 95, 97, 99, 112, 118
Peroxymonosulfate, 79
Persano, L., 53
Persson, C., 27
Pesquera, D., 84
Petroleum, 44, 110
PFM, 77, 93, 96
Pharmaceutical, 18, 55
Phase, 1, 3, 9–13, 17, 22, 23, 34–36, 45, 46, 48, 49, 62, 63, 90, 91, 93, 95, 96, 101–107, 110–112, 115, 118, 119
  phase-transformation, 110
  phase–voltage, 96
Phenol\phenolic, 18, 43, 79
Philip, J., 99
Phonon, 13, 37, 46, 105, 106, 108, 111
Phosphorus, 71
Photo, 8
Photoactivity, 23, 43
Photoanode, 43, 44
Photocatalysis, 6, 13, 21, 23, 55, 56, 58, 61, 90
Photocatalysts, 4, 8, 11, 18, 22, 23, 44, 55–58, 61, 62, 77, 78, 80, 103, 110, 112, 117, 118
Photocathode, 44
Photochemical, 112
Photochromic, 4, 8, 15
Photochromism, 6, 9
Photoconductive, 71
Photo-degradation, 18
Photodetectors, 9
Photoelectric, 107
Photoelectrochemical, 4, 25, 43, 44
Photo-excited, 55, 56
Photogenerated, 23, 58, 61
Photoluminescence, 4, 9, 72
Photonics, 13
Photons, 105
Photoreponsivity, 71, 118
Photoresponsivity, 71
Photosensitivity, 77
Photovolta-chromic, 12
Photovoltaics, 1, 43, 48, 77, 117
Phuruangrat, A., 28
Physicochemical, 55
Physics, 3, 71, 94, 105, 107
Physisorption, 44
Piccirillo, C., 113
Piezoelectricity, 48, 77, 90, 93, 95, 96
Piezo-voltage, 48
Piloto, C., 52
Pimentel, J. A. I., 30
Pincus, P. 116
Pinheiro, D., 31
Pinned, 35
Pinning, 35, 93
Pino, F., 5
Pisignano, D., 53
Pitch, 43
Pitts, 6, 28
Planar, 37, 76, 78
Planes, 2, 8, 35, 37, 49, 79, 108
Plashnitsa, V. V., 25
Plasma, 62
Plasma-treated, 61, 62
Plasmonics, 44
Plastic, 65
Platelets, 10, 110
PLD, 91, 95, 104
Plenum, 51
Plummer, E. W., 5, 99
Plum-pudding-like, 20
PMN-PT, 76, 77
p-n, 48, 56, 80
p-NiO, 58
Poilblanc, D., 50
Polar, 2, 82, 95
Polariton, 13
Polarity, 2, 117
Polarizability, 2, 3
Polarization, 13, 72, 73, 76, 77, 90, 92, 95, 96, 106
  polarization-dependent, 72
  polarization-induced, 73, 76
Polaronic, 3, 37, 40
Polarz, S., 83
Poli, N., 32
Pollutants, 10, 18, 55, 73, 110
Polyaniline, 65
Polycrystallinity, 44, 90, 91
Polymer, 20, 52, 65, 90, 92
  polymer-clothed, 18
  polymeric, 93
Polymorphs, 12, 13, 40, 44, 49, 62

Polypyrrole, 65
Poly-sulphide, 82
Pomar, A., 99, 100
Ponnusamy, D., 30
Ponpandian, N., 29
Poortmans, J., 28
Popovic, S., 32
Porada, S., 83
Pore, 64, 69
Porosity, 21
Porous, 11, 20, 42, 45, 63, 66, 69, 79, 80, 91, 107, 108
  porous-graphene-wrapped, 79
Porta, F. A. La, 115
Portable, 20
Portnoi, M., 114
Posadas, A., 84
Post-annealing, 22
Post-oxidation, 104
Poswal, H. K., 113
Poteryaev, A., 112
Pouget, J. P., 112
Povey, I. M., 84
Powell, M. J., 114
ppb, 12, 45, 46, 49, 50, 59, 65, 66
  ppb-level, 66
PPLN, 72, 77
ppm, 12, 15, 18, 19, 45, 46, 48–50, 59, 65–67, 72, 73
  ppm-level, 66
p-PPy, 67
  PPy, 65, 67
Pradhan, N. R., 83
Prasad, A. K., 29, 30, 52–54, 68, 108, 112, 115
Prashanth, M. K., 30
Pratsinis, S. E., 70
Pre-adsorbed, 44
Presser, 83
Pressure, 1, 2, 104, 105, 108, 110, 115, 119
  pressure-dependent, 110
  pressure-induced, 105, 110, 119
  pressure-transmitting, 110
Pristine, 19, 40, 43, 49, 50, 61, 64, 79
Prontera, C. T., 27
Propylene, 65
Protonation, 67
Protonic, 22
Protons, 67
Pseudo, 34, 35
  pseudocapacitive, 62, 65
  pseudocapacitor, 80
  pseudogap, 36, 38, 117
Pt-activated, 49
Pt-assisted, 118
Pt-decorated, 15
Pt-functionalized, 45
Pt-graphene, 77
PTO, 90, 91, 95
p-type, 44–46, 48, 62, 66, 73, 79, 80
Pu, 28, 68, 116
Pugliese, M., 12, 27
Puig, T., 99
Pulsed, 15, 73, 104, 105
  pulsed-laser, 91
Pulverization, 20, 40, 78, 79
Puma, G. L., 52
Pumice-TiO, 22
Purushothaman, K. K., 87
Pushpan, S., 29
Putthanarat, S., 52
PVA, 108
PVD, 103, 104
Pyramids, 15, 16
Pyroelectricity, 90
Pyrolysis, 18
Pytte, E., 116
PZO, 91
PZT, 72, 76, 90, 91, 93, 94

Q

QDs, 55–59, 61, 78, 81, 118
  QD-based, 59
  QD-decorated, 58
Qian, G. Y., 113
Qian, T., 97, 98
Qian, Y. T., 28
QianYi, C., 86
Qiao, S. Z., 86
Qiao, S.-Z., 25, 67
Qiao, Y. Q., 87
Qi, D.-C., 29
Qi, J., 51
Qi, L., 86
Qin, F., 27
Qing, L., 86
Qin, J. J., 67
Qin, K., 30
Qin, W., 51
Qin, Y., 69
Qin, Z., 29, 68
Qi, Q., 88
Qi, R., 88
Qiu, C. W., 28
Qiu, J., 28, 86
Qiu, L., 52
Qiu, S. Y., 26
Qiu, X., 31
Qiu, Y., 30
Qiu, Z., 69, 88

Qi, Y., 16, 30
Qi, Z., 83
Quan, F., 86
Quantization, 74, 75
Quantum, 2, 4, 8, 23, 40, 50, 55, 74, 76, 78, 93, 95, 109, 110, 116, 118
Quasi-elastic, 83, 111
Quasi-industrial, 20
Quasi-metallic, 14
Quasi-one-dimensional, 34
Quasi-solid, 43
Qu, B., 86
Qu, J., 86
Qu, Z., 53, 86
Qu, W., 100

R

Rabani, I., 30
Rad, T. S., 22, 32
Radenovic, A., 83
Radhakrishnan, R. P., 16, 18, 30
Radiation, 4, 61, 107, 110
Radicals, 55, 56, 58, 61, 62
Radio, 104
Radisavljevic, B., 83
Radius, 81
Raghavendra, K. G., 113
Raghu, M. S., 30
Rahimi, F., 27
Rahman, F., 29
Rahman, M. A., 29
Rahman, Q. I., 30
Rahmani, M. B., 29
Raible, I., 29, 53
Raj, C. J., 51
Rajamathi, C. R., 85, 112
Rajamathi, M., 85
Rajapitamahuni, A., 84, 85
Rajarajan, M., 29, 115
Rajeswaran, B., 112
Rakhi, R. B., 89
Ráksi, F., 114
Raman, 10, 11, 13, 14, 16, 17, 22, 27, 59, 60, 77, 105, 106, 110, 111
Ramanathan, S., 114, 115
Ramesh, R., 98–100
Ramgir, N. S., 69, 70
Ramirez, J. G., 114
Ramirez, R., 30
Ramkrishna, S., 97
Ramos Ramón, J. S., 29
Ramos, C. A., 99
Ramsay, L., 98
Ran, J., 67
Ran, Y., 54
Rani, M., 6
Ranjbar, S., 28
Rao, C. N. R., 5
Rao, K. V., 5
Rao, P. M., 29
Rao, S. S., 97
Rasing, T., 98
Rasool, K., 99
Rasool, H. I., 84
Rastelli, A., 85
Ratan, A., 5
Ratio, 8, 23, 41, 55, 60, 61, 66, 77, 80, 91, 92, 94, 96, 106, 111
Rattanasakuldilok, W., 87
Ravelosona, D., 113
Ravichandran, J., 84
Ravichandran, K., 67
Ravindran, T. R., 113
Ravishankar, T. N., 25, 33
Ray, A., 68
Rayappan, J. B. B., 69
Raychaudhuri, A. K., 97
Raychaudhury, M. D., 114
Raza, R., 6
Razi, F., 27
Razzaq, H., 53
Reaction, 43, 44, 46, 49, 59, 61–64, 77–79, 82
Reactivity, 23, 42, 58, 61, 95
Real-time, 108
Rear-view, 20
Rechargeable, 4, 6, 71, 78
Reciprocal, 36
Recombination, 1, 23, 58
Reconfigurable, 76
Reconstruction, 7
Recovery, 12, 19, 45, 48, 49, 59, 65, 66, 73
Rectification, 93
Recyclable, 56
Red, 38
Reddeppad, M., 26
Reddy, B. M., 31
Reddy, G. B., 28
Reddy, V. R. Minnam, 26
Reddy, Y. A. Kumar, 26
Redox, 13, 23, 56, 59, 61–63, 65, 74
Reduction, 1–2, 5, 8, 13, 14, 17, 20, 23, 25, 44, 56, 58, 59, 61, 63, 65, 66, 72, 74, 75, 78–80, 90, 93, 103–105, 107
Refining, 21
Reflectance, 11
Reflection, 107
Refractive, 12
Reghunath, S., 31
Regier, T., 86

Reichelt, W., 112
Relaxation, 77
Remita, H., 32
Remy, C., 67
Reneker, D., 52
Renewable, 56, 62, 78
Ren, H., 51, 68, 86
Ren, J., 69, 70, 88, 116
Ren, T., 85
Ren, W., 32, 100
Ren, Y., 85, 98
Ren, Z., 32
Ren, Z. F., 97
Ren, Z. H., 97
ReO, 2, 49
Repelin, Y., 30
Reshma, P. R., 30, 68, 115
Reshmi, R., 29
Resistance, 18, 19, 44, 45, 48, 107, 108
Resistivity, 15, 37, 65, 73, 80, 105, 118
Resmini, A., 54
Resolution, 23
Resonance, 92
Response, 4, 12, 15, 18, 19, 23, 43, 45–50, 59, 65–67, 72, 73, 108
Retention, 20, 40–43, 59, 63, 64, 73, 108
Revcolevschi, A., 115, 116
RF, 13, 104
  RF-sputtered, 104
rGO, 20, 58, 63, 67, 112
RhB, 56, 58, 62
Rhodamine, 56, 78
Rhodes, D., 83
Rhombohedral, 21, 44
Ribbons, 79
Rice, 26, 50, 103, 112
Richardella, A., 83
Richardson, M. A., 113
Ridley, I., 113
Righi, T. M., 53
Rios, S., 100
Rischau, C. W., 83
Rizzo, A. 27
RMO, 4
Roach, W. R., 105, 114
Robert, D., 32
Robinson, D. B., 33
Robinson, J. T., 87
Robinson, R. D., 25
Rockenberger, 98
Rocksalt, 2
Rocksalt-type, 2
Rodin, V. N., 26
Rodríguez, L. A., 99
Rogers, S. P., 85
Rohrer, G. S., 28
Rondinelli, J. M., 83
Rong, Z., 30
Roppolo, M., 28
Roro, K., 53
Rosenberg, E. R., 114
Rosenbusch, M., 99
Ross, C., 114
Rosseinsky, D., 6
Rotations, 16, 37
Rothschild, A., 53
Rougier, A., 33
Rovida, G., 113
Roy, A., 68
Roytburd, A. L., 100
Rozenberg, M. J., 113
RSC, 27, 30, 33, 68, 69, 84, 87, 89
RT, 59, 65–67
Ru, 66, 94
Ruan, G., 26
Ruan, H., 26
Ruan, M., 29
Ruan, S., 69
Ruddelsden–Popper-type, 2, 35, 117
Rui, X., 16, 30, 69, 85
Ruiz-Hitzky, E., 30
Rumo, M., 116
Ruoff, R. S., 25, 26, 86
Russo, S. P., 25
Rutile, 2, 21–23, 48, 58, 101, 102, 106
Rutledge, G. C., 53
Rørvik, P. M., 98

S

Sacanell, J. 99
Saccomanni, A., 32
Sachan, V., 5, 50
Sachs, R., 84
Sacrificial, 42, 95
Sadale, S. B., 27
Sadishkumar, V., 30
Sae, J., 5
Saeki, H., 99
Safi, T. S., 105, 114
Saghatforoush, L. A., 27
Saghayezhian, M., 5
Saha, S., 68, 84
Sahoo, P., 52
Sahu, N. K., 6
Saint-Martin, R., 116
Saito, K., 99
Saitoh, E., 116
Saitzek, S., 104, 113
Sakhuja, N., 15, 29

Salahuddin, S., 25, 114
Salam, K. N., 52
Salce, B., 111, 116
Salih, E., 5
Salmeron, M., 115
Salt, 40, 90
Saltwater, 58
Salunkhe, R. R., 31
Samadpour, M., 5
Samant, M. G., 114
Samanta, S., 97
Samarth, N., 83
Sambi, M., 113
Sambrano, J. R., 115
San, X., 30
Sanal, K. C., 29
Sanati, S., 27
Sánchez, F., 84
Sánchez, R. D., 99
Sander, B. M. S., 32
Sandiumenge, F., 99, 100
Sandu, C. S., 97
Sandwich, 77
  sandwich-structured, 77
Sang, D., 26
Sanjeeviraja, C., 28
Sankaranarayanan, S., 6
Sapphire, 104
Sarangi, S., 97
Saravanakumar, B., 87
Sardar, K., 69
Sardar, M., 112, 114
Sarkar, S., 69
Sarro, P. M., 26
Sasago, Y., 116
Sasaki, J. M., 27
Sasaki, T., 25, 26
Sastri, M. L. S., 30
Satellite, 37
Sato, F., 27
Sato, H., 26
Sato, M., 114, 116
Saturation, 76, 74–76
Sauer, J., 30, 113
Savu, R., 87
Sawa, A., 84
SBT, 94
SBTO, 91
Sc, 4
Scale, 20, 93, 117
Scancella, M. T., 98
Scandium, 4
Scattering, 37, 93, 110
Schaak, R. E., 26
Schalch, J. S., 107, 114
Scheer, E., 83
Scher, E. C., 98
Schladt, T. D., 114
Schläefer, J., 114
Schlecht, U., 29, 53
Schlotter, W., 116
Schmehl, A., 84
Schmidt, M., 84
Schmidt, O. G., 85
Schmidt, V., 98
Schmitt, T., 116
Schmuki, P., 33
Schnabel, T., 28
Schneider, C. W., 84
Schuller, I., 5
Schuller, I. K., 113, 114
Schulz, H. L., 50
Schulze, S., 98
Schwarz, U. 115, 116
Scigaj, M., 84
Scott, J. F., 98–100
SCs, 59, 62, 63
Seabaugh, A., 26
Seawater, 82
Second-order, 34, 105, 118
Sediri, F., 113
Sekar, S., 18, 30
Selectivity, 12, 19, 46, 48, 49, 65–67, 72, 79, 108
Self-aggregation, 20, 40
Self-assembly, 40, 81
Self-discharge, 19
Self-powered, 48
Self-trenching, 94
Selli, E., 31, 32
SEM, 47, 57
Semiconducting, 12, 14, 22, 23, 45, 46, 48, 55, 64, 72, 97, 108
Semiconductor, 2, 7, 8, 14, 15, 23–25, 32, 44–46, 48, 49, 55, 71, 72, 79, 80, 93, 107
  semiconductor-based, 44
Sengupta, S., 114
Senjyu, T., 5
Sensing, 8, 10, 12, 15, 18, 19, 22, 44–46, 48–50, 59, 61, 65–67, 72, 73, 80, 108, 118
Sensitivity, 15, 44–46, 48, 66, 67, 72, 73, 108
Sensors, 1, 6, 8, 11–13, 15, 17, 18, 29, 32, 44, 47–49, 54, 65–67, 72, 73, 76, 79, 80, 90, 101, 108, 109
Senz, S., 98
Seo, B. I., 98
Seo, D.-H., 31, 51, 84, 86
Seo, S., 84
Seo, Y.-S., 30
Seong, M. J., 28
Separation, 23, 55, 58, 61, 66, 93, 105, 106

Seriwatanachai, C., 87
SERS, 59, 60
Sethuraman, B., 87
Sethuraman, K., 30
Setter, N., 97
Sha, J., 97, 99
Shafiei, M., 52, 53
Shaikhutdinov, S., 30, 113
Shaislamov, U. A. 98
Shang, D. S., 25
Shan, J., 25
Shankar, K. S., 97
Shanker, U., 6
Shanmugam, M., 30
Shan, X.-Y., 86
Shao, C. L., 26
Shao, G., 86
Shaoyan, W., 114
Shaposhnikov, V. L., 26
Sharma, P., 83, 85
Sharma, S., 115
Sharma, S. K., 115
Sharma, Y., 86
Shchukarev, A., 53
Shear, 16, 49
Sheets, 78, 80, 95
Shejwal, R. V., 69
Shell, 20, 42, 91
Shelton, R. N., 50
Shen, B., 28
Shen, D. Z., 26
Shen, F., 32
Shen, G., 97
Sheng, L., 32, 83
Shen, H., 27, 84
Shen, J., 98, 99
Shen, J.-Y., 12, 28, 70
Shen, L. M., 32, 97
Shenoy, V. B., 88
Shen, Q., 86
Shen, S., 29
Shen, W., 68
Shen, X., 25
Shen, Y., 30
Shen, Z., 26, 52
Shen, Z.-X., 31, 85
Sherstyuk, N. E., 98
Shibuya, K., 114
Shieh, S. R., 116
Shi, F., 86
Shift, 9, 22, 23, 46, 56, 58, 75, 77, 92, 105, 106
Shi, G., 25
Shi, J., 84, 87, 114
Shi, K., 52
Shi, L., 25, 86
Shim, M., 85
Shimne, G., 5, 50
Shim, Y. S., 27
Shinde, S. R., 99
Shin, H. S., 85
Shiomi, Y., 116
Shi, Q., 27
Shi, S. J., 68, 87
Shi, W., 68
Shi, Y. M., 83, 98
Shklovskii, B. I., 23, 32
Shokouhimehr, M., 68
Short-lived, 55
Short-range, 111
Shruthi, J., 29
Shu, X., 32, 70
Shu, Y., 86
Shyue, J. J., 98
Si, 10, 20, 28, 30, 60, 77, 84, 93, 94, 97, 104, 111
  Si-based, 72, 74
  Si-doped, 66
Sian, T. S., 13, 28
SIBs, 78, 81
Siciliano, P., 68, 69
Siddaramanna, A., 113
Siddiqui, M. J., 68
Siders, C. W., 114
Sievi, N. A., 70
Sigov, A. S., 98, 100
Silencing, 4
Silica, 65
Silicene, 7
Silicon, 96
Siligardi, C., 31
Sim, D., 85
Simon, P., 25
Sinaim, H., 28
Singapore, 83
Singh, A., 25
Singh, B., 87
Singh-Bhalla, G., 99
Singh, I., 5
Singh, R. K., 87
Singh, S., 98
Singh, V. R., 5, 107, 114
Single, 7, 41, 46, 77, 91–93, 102, 103, 105, 109, 111
  single-crystalline, 74, 76, 90, 109, 110
Single-layered, 22
Singlets, 111
Sinha, S. K., 30, 68
Sinitskii, A., 83–85
SiO, 72, 77, 92
Si-processing, 72
Sisman, O., 22, 32, 83

Sisson, D. L., 116
Sivakumar, P., 51
Sivakumar, R., 13, 28
Size-dependent, 55, 90
Slater, J. C., 3, 5, 113
Smart-window, 107
Smirnov, M., 30
Smirnov, V. M., 113
Smith, K. E., 114
Smith, R. L., 28
Smith, T. J., 116
Sn, 49
SnO, 23, 42, 45, 46, 48, 49, 63, 72, 73, 81, 104, 107
Sodium, 66
Sohma, A., 116
Sohn, W., 27
Sol, 91
Solar, 13, 21, 22, 25, 43, 56, 58, 62, 71, 80, 107
  solar-charged, 43
Solgel, 8, 10, 13, 15, 22, 25, 56, 90, 91, 93, 98, 103, 104
  solgel-assisted, 22
Soliton, 106
Solution, 60, 61, 79, 92, 112
  solution-based, 95
  solution-phase, 90
Solvothermal, 22, 59, 62, 66, 90
Solyom, J., 113
Solzbacher, F., 28
Sommers, C., 103, 112
Song, C., 86, 88
Song, E. B., 84
Song, F., 52
Song, H. C., 52, 54, 84, 100
Song, J. M., 54, 87, 98, 100
Song, L., 68
Song, P., 29, 84
Song, Q., 31
Song, R. Q., 109, 115
Song, S., 115
Song, W., 10, 27
Song, X., 33, 51
Song, Y., 26, 84
Song, Z., 68
Son, J., 115
Sonocatalytic, 22
Son, Y.-D., 67
Son, Y. R., 6
Sood, S., 67
Sorption, 46, 61
Sousa, C. T., 99
Souza Filho, A. G., 11, 27
Sow, C. H., 99
Spadaro, D., 31
Spanier, J. E., 97
Spectroscopy, 10, 11, 13, 16, 22, 35, 56, 59, 77
Spencer, M. G., 25
Spin, 3, 5, 13, 16, 34–37, 71, 91, 93, 101, 103–107, 109, 111, 117, 118
Spinel, 41, 46, 59, 118
Spinon, 106, 107, 111, 118
Spin-peierls, 103, 108, 110, 111, 118
Spin-soliton, 111
Spintronic, 90, 93
Spizziri, P. G., 53
Splitting, 3, 9, 43, 44, 78, 95
Spray, 18
Springer, 6, 10, 32
Spronken, G., 113
SPT, 101, 103–105, 107, 118
Sputter-deposited, 58
Sputtering, 10, 13, 16, 18, 19, 91, 104
Squier, J. A., 114
Srathongsian, L., 87
Srihari, V., 113
Srikanth, H., 97
Srinivasan, G., 97
Srinivasan, M., 86
Srinivasan, P., 69
Srinivasan, V., 115
Sriram, S., 28–30
Srivastava, A., 84
Srivastava, S. K., 113
Srivastava, V. K., 83
SrRuO, 72
SrTiO, 71, 72, 74–76, 90
SrVO, 95
S-scheme, 61
Stability, 2, 4, 7, 13–15, 18–22, 40, 42, 43, 45, 48, 49, 55, 56, 59, 63, 65, 73, 78–81, 95, 108, 118
Stadler, F. J., 69, 70
Staiger, M. P., 53
Staikov, G., 84
Stalick, J. K., 116
Stan, L., 83
Stanciu, L., 85
Stanger, J. J., 53
Stavila, V., 33
Steeber, D. A., 87
Stefanovich, D., 106, 114
Stefanovich, G., 106, 114
Steffes, H., 28
Steinhart, M., 98, 100
Stephens, Z. D., 33
Step-scheme, 61
Sternlleb, B. J., 5, 50
STM, 35, 38
STO, 90–92, 94, 95

Stokes, 106
Stoliar, P., 99
Stoller, M. D., 26, 86
Storage, 1, 4, 6, 8, 19, 20, 40, 42, 43, 46, 51, 58, 59, 62, 63, 71, 78–82, 108, 111, 118
Strain, 1, 9, 77, 79, 80, 93, 104
Strano, M. S., 26
Strelcov, E., 112
Streltsov, S. V., 112
Stress, 108
Stripes, 35, 36, 117
Strocov, V. N., 116
Structure, 1–3, 7–15, 17–23, 25, 27, 29, 31, 33, 35–39, 43–46, 49, 51, 60, 62, 63, 71–73, 77–80, 82, 94, 95, 101–111, 117, 118
  structure-property, 1
Strunk, J., 29
Stucky, G. D., 53
Sub-bands, 3
Sublattice, 101
Sub-ppm, 15
Substrate, 13, 42, 47, 60, 65, 72, 74–78, 80, 91–97, 104, 108
Suenaga, M., 50
Sugahara, T., 69
Suganthi, A., 29, 115
Suganuma, K., 28, 69
Suh, D., 83, 84
Suh, J., 114
Suh, K. S., 85
Sui, D., 86
Sui, L.-L., 29
Sukang, B., 85
Sulaiman, Y., 69
Sulphate, 22
Sulphur, 81
Sum, T. C., 31
Sumboja, A., 51
Sun, B., 87
Sun, C.-J., 5, 85
Sun, C.-L., 87
Sung, H. J., 51
Sun, G. J., 52–54
Sungpanich, J., 53
Sun, H., 52, 68, 87, 97
Sun, H.-B., 83
Sun, H. Y., 100
Sun, J. L., 13, 26, 28, 70, 84, 87, 100
Sun, K. N., 87, 88
Sun, L., 70
Sunlight, 18, 56, 58, 112
Sun, N., 83, 88
Sun, P., 54
Sun, Q., 69
Sun, R., 27
Sun, S., 84, 87, 88
Sun, X. C., 5, 31, 51, 53, 69, 88, 97
Sun, Y.-H., 25, 27, 31, 54, 83, 86, 87
Sun, Y. L., 85
Sun, Z. H., 5, 31, 32, 88, 98
Supercapacitors, 1, 4, 41, 42, 59, 78–80, 82, 108, 118
Superconductivity, 1, 34, 35, 50, 71, 90, 117
Superconductor, 2, 34, 35, 38, 117
Super-exchange, 3, 4
Superlattices, 36, 71, 76, 97
Superoxides, 55
Superparamagnetism, 93
Suresh, S., 5
Surface, 1, 2, 4, 5, 7–9, 11, 12, 15, 17–22, 36, 37, 40, 41, 43–46, 48, 49, 55, 56, 58, 61–63, 65–67, 71–76, 78–81, 92, 93, 95, 104, 107, 117, 118
  surface-depletion, 45
  surface-engineered, 40
  surface-enhanced, 59
  surface-resistance-control, 45
  surface-to-volume, 8, 23, 55, 61
Surfactant, 22
  surfactant-assisted, 109
  surfactant-coated, 59
Surya, S. G., 52
Susceptibility, 105, 111
Suzuki, V. Y., 110, 112, 115
Swart, H. C., 53
Switchable, 90, 96
Switches, 8, 15, 17, 21, 71–73, 76, 77, 79, 80, 94, 101, 107
Syassen, K., 115, 116
Symmetry, 22, 34, 79, 110, 111
Synchrotron, 110
Synergistic, 14, 61, 62, 77, 79, 81, 104
Synthesis, 8–10, 12, 13, 15, 16, 18–20, 22, 40–42, 56, 58, 59, 91, 95, 103, 104, 109, 110, 112
Syringes, 95
System, 2, 10, 14, 24, 40, 58, 62, 67, 80, 104, 108, 110, 111
Szafraniak, I., 98, 100
Szekeres, A., 28
Szot, K., 84

T

T, 101, 105, 106
Tabata, H., 99
Tachibana, Y., 25
Tada, H., 33
Tadanaga, K., 98
Ta-doped, 25
Tafel, 43

Taga, Y., 32
Tai, G., 31
Takada, K., 26
Takagi, H., 50
Takahashi, K., 84
Takano, M., 50
Takeuchi, T., 50
Takeya, J., 111, 116
Talanta, 70
Tanaka, A., 112
Tanaka, I., 116
Tan, C., 26
Tang, D., 69
Tang, H., 51, 116
Tang, K., 28
Tang, M. H., 84
Tang, N., 67, 89
Tang, W. H., 89, 99
Tang, X. G., 68, 98
Tang, Y. L., 27, 31, 84, 85, 114, 116, 100
Tang, Z., 51
Tan, H. T., 69
Tan, Q., 28
Tan, W., 115
Tan, X., 28
Tan, Z., 33
Tao, H. Z., 113
Tao, W., 86
Tao, X. Q., 27
Tarascon, J. M., 31
Target, 10, 18, 104
Tari, S., 99
Tatarenko, A., 97
Tatsumisago, M., 98
Tatsuta, T., 100
Tay, B. K., 5
Tay, Q., 31
Taychatanapat, T., 87
Taylor, S., 88
tBuN, 10
TC, 18, 61, 62
TEA, 65, 67
Technique, 8, 10, 13, 16, 18, 22, 42, 44, 48, 56, 58, 59, 66, 72, 73, 82, 90–92, 95, 103, 104
Technologies, 4, 6, 13, 20, 44, 78, 103, 115
Teh, C. M., 23, 32
Tehrani, F. S., 69
Teja, M., 52
TEM, 41, 57
Temperatures, 1–4, 9–13, 15, 17–19, 22, 23, 34, 35, 45–50, 59, 65, 66, 72, 73, 75, 76, 78, 80, 91, 93, 94, 101, 103–109, 111
  temperature-dependent, 37
  temperature-driven, 40
Templates, 42, 90, 91
  template-assisted, 22, 90–93
  template-based, 20
  template-grown, 93
Teng, Y., 69
Teo, W. E., 97
Terabit, 92
Terahertz, 13, 93, 94, 107
Terasaki, I., 115
Teresa, J. M. De, 99
Terminal, 13, 15, 95
Termination, 80
Ternary, 58
Terrones, H., 83
Terrones, M., 25
TERS, 105
Tesfamichael, T., 29, 53
Tetracycline, 18, 61
Tetragonal, 9, 21, 22, 35, 90, 95, 101, 102, 110
Tetrahedra, 14, 108
Textile, 55
Thakur, S., 69
Thakur, V. K., 33
Thambidurai, S., 5
Thangadurai, P., 30
Theatrically, 13
Theodore, S. J., 70
Theoretical, 13, 16, 59, 62, 65, 78, 81, 95
Theory, 2, 3, 22, 43, 50, 65, 95, 103
Thermal, 4, 10, 13, 18, 23, 34, 45, 56, 80, 101, 105, 106, 108, 111
  Thermal-assisted, 73
Thermochromic, 4, 17, 78, 107
Thermodynamic, 12, 14, 34
Thermoelectric, 105, 111
Thick, 11, 13, 104
Thicknesses, 8, 18–20, 22, 58, 74, 76, 90–93, 95, 96
  thickness-controlled, 10
  thickness-dependent, 22, 72
Thiel, S., 84
Thin, 4, 6, 10, 12–15, 18, 19, 23, 27–29, 32, 43, 54, 66, 72, 74–76, 80, 92, 93, 99, 100, 103, 104, 107, 108, 113
  thin-film, 74, 76
Thind, S. S., 10, 26, 27
Thiyagarajan, P., 31
Thomas, T., 15, 29
Thongtem, S., 28, 53
Thongtem, T., 28, 53
Thornton, M. J., 5
Three-band, 103
Three-dimensional, 78
Three-order, 59, 93
Threshold, 74, 107

Thwala, M. M., 26
Ti, 1, 4, 21, 25, 43, 72, 73, 76, 77, 81, 82, 90, 91, 94
Tian, B. B., 84
Tian, C., 32
Tian, G., 32
Tian, H., 85
Tian, J., 68
Tian, K., 54
Tian, L.-L., 86
Tian, M., 27
Tian, Q., 53, 54
Tian, W., 88, 114
Tian, X., 69, 85
Tian, Y., 68, 70
Tian, Z. R., 32
Tien, H. N., 87
Tilt, 37, 101, 102, 104
Ting, Y., 85
Ti-O, 22
TiO_2, 2, 4, 7, 18, 21–23, 25, 43, 44, 48, 49, 56–59, 61, 62, 64–66, 72, 73, 76–78, 81, 82, 91, 95, 104, 107, 117, 118
TiO_6, 22
Ti-O-Ti, 22
Tip, 79, 92, 94
Tischler, J. Z., 112
Titanate, 22
Titania, 77
Titanium, 4, 22, 48
Titanyl, 22
Tiwary, C. S., 25
Tjeng, L. H., 112
Tjiu, W. W., 53
T-like, 96
TMA, 19, 49
TMDs, 64, 71, 72, 90, 97, 118
TMOs, 1–5, 7–9, 11–17, 19, 21, 23, 25, 27, 29, 31, 33, 34, 37, 40, 42–44, 48, 55, 59, 61–65, 71–83, 85, 87, 89, 117–119
  TMO-activated, 63
  TMO-based, 1, 7, 62, 76, 118
Tokunaga, M., 115
Tokura, Y., 6, 83, 114
Tolerance, 66
Toluene, 15, 46, 50, 66
Tomar, M., 5
Tong, P. V., 53
Tong, T., 68
Tong, W., 69
Tong, Y., 51
Tong, Z., 70
Topography, 35, 73, 92
Toroidal, 95, 97
Torr, 44
Torrance, J. B., 6
Torrini, M. 113
Tóth, Cs., 114
Touati, F., 113
Toxicity, 56, 62
TQDs, 57
Trace, 46, 65
Tranquada, J. M., 5, 50
Transducers, 8
Transfer, 2, 4, 11, 14, 24, 40, 44, 56, 59, 62, 63, 65–67, 72, 79, 80, 95, 108
  transfer-induced, 59
Transformation, 4, 22, 35, 38, 63, 79, 95, 110
Transient, 47
Transistors, 8, 72, 76, 107
Transitions, 1–12, 14, 16–18, 20–22, 24–26, 28, 30, 32, 34–72, 74, 76–78, 80, 82, 84, 86, 88, 90–96, 98, 100–119
Transparency, 8, 20, 43, 58, 107
Transporting, 8, 41–43, 71, 74, 80–82, 93, 95, 111, 118
Trapping, 56, 74, 75–77
Trastoy, J., 113
Travaglini, G., 50
Treciakowski, W. A., 115
Triclinic, 9–11, 63, 101
  triclinic-WO, 10
Triethylamine, 65
Trimethylamine, 15, 19, 49
Trions, 72
Trioxides, 49, 54
Tripathi, M. 32
Triplet, 111
Triplon, 111
Triply, 3, 13
Tripp, C. P., 32
Triscone, J.-M., 83
Trithoxysilane, 77
Trodahl, H. J., 97
Troiani, H. E., 99
Trung, D. D., 53
Tryk, D. A., 32
Tsang, Y., 68
Tselev, A., 112
Tseng, Y., 116
Tserkovnyak, Y., 114
Tseung, C. C., 5
Tsuchiya, K., 30
Tsuji, O., 100
Tsukada, I., 116
Tsymbal, E. Y., 83, 84, 97
Tucker, N., 53
Tufail, A., 67
Tu, J. P., 51, 87
Tuller, H. L., 53
Tunability, 55
Tunable, 21, 77, 80

Tungstate, 66
Tungsten, 49, 54
Tunings, 8, 20
Tunnell, 34, 35, 72, 73, 77, 94
Turkovic, A., 32
Turner, J. A. 6
Tu, W., 85
Tu, Y., 29
Twin, 95
Twist, 43, 101
Two-dimensional, 7, 11, 18, 20, 63, 71
Two-step, 49
Two-terminal, 77, 106
Tyagi, A. K., 52, 53, 112
Type-II, 56, 58

U

U, 2, 3, 58, 103
Uchida, S., 50
Uchinokura, K., 115, 116
Uchlda, S., 5, 50
Udagawa, M., 116
Ultra, 56
Ultrafast, 105
Ultrahigh, 81
Ultra-large, 40
Ultra-sensitive, 90
Ultrasmall, 65
Ultrathin, 4, 20, 23, 40, 65, 78, 95, 96, 117
ultrathin-derived, 79
Umar, A., 30, 67
Umarji, A. M., 112
Uncoated, 48
Uncompensated, 93
Undecorated, 59, 65, 66
Under-doped, 39
Undoped, 66
Uneven-uneven, 35
Unidirectional, 118
Uniform, 10, 15
Unit-cell, 96
Universality, 105
Unsaturation, 2, 7, 55, 117, 118
Unshared, 13
Unstable, 34, 108
Urban, J. J., 44, 97, 99, 114
Urbanowicz, A., 99
USA, 68
Utrera, L., 30
UV, 4, 18, 56, 61, 110, 112

V

V, 1, 4, 102, 105
Vacancies, 2, 12, 14, 17, 43, 45, 66, 74, 76, 80, 104, 105
Vacant, 45
Vacuum, 29, 53, 54
Valencies, 2, 11, 23, 25, 35, 58, 59, 77, 91, 95
Valentin, C. D., 27
Valle, 113
Valmianski, 114
Values, 3, 35, 41–44, 48–50, 59, 61–63, 65–67, 78, 80, 82, 93, 103, 104, 107, 110, 118
Valvidares, S. M., 84
van, 7, 8, 33, 46, 62, 68, 71, 72, 81, 83, 115, 117
Vanadates, 20
Vanadium, 4, 15, 17, 18, 20, 40, 44, 46, 101, 104, 105, 112, 118
Vanadyl, 15, 18
van der Marel, D., 115
van der Wal, A., 83
Vapours, 10, 13, 15, 22, 46, 77, 103
Varela, M., 84
Varga, T., 27
Variable, 23, 35
Variations\variating, 1, 2, 13, 20, 25, 42, 43, 62, 65, 72, 76, 77, 107
Varshney, B., 68
Vasileff, A., 25
Vasiliev, A. L., 98
Vasiljev, V. A., 98
VC, 81
VCH, 6
vdW-based, 7, 8, 13, 117
vdW-bonded, 8, 12, 13, 15–17, 117
vdW-type, 8
Vectors\vectorial, 24, 35, 38, 39
Vegetables, 79
Venkatachalam, M., 5
Venkataramana, B., 52
Venkatesan, T., 84
Vennik, J., 30
Verma, N., 5
Vermang, B., 28
Vernardou, D., 103, 113
Vershinin, M., 50
Vertical, 35, 39, 77, 79
Vertices, 15
Verykios, X. E., 23, 32
Ves, S., 115
VI, 58
Vibrations, 10, 11, 13, 16, 17, 22, 59, 106, 107, 119
Victoria, 18
Vijjapu, M., 52
Villegas, J. E., 113
Villeneuve, G., 112
Villevieille, C., 28
Vinesh, V., 30
Violet, 18

Visible, 4, 6, 11, 18, 22, 23, 43, 44, 56, 58, 61, 62, 72, 73, 77, 105
visible-light, 58
Vital, 104
Vitamin, 58
Vittoria, M., 31
Vivekanandhan, S., 6
Vlooswijk, A. H. G., 100
V-O, 15–17
Vo, T. H., 85
VO_2, 2, 17, 20, 40, 46, 47, 63, 78, 80, 82, 101–108, 118, 119
VO_5, 15, 16
$V2O_5$, 4, 7–9, 15–21, 40–42, 46, 62, 63, 74, 104, 105, 117, 118
$V2O_3$, 2, 17, 20, 63, 82, 105
VOCs, 44–46, 65
Voevodin, A. A., 83
Vogt, T., 26
Voigt, J. A., 32
Volanti, D. P., 27
Volatile, 44, 65, 95
Vollenkle, H., 115
Voltage, 35, 38, 39, 73–76, 91, 93, 106, 108
Voltage-sweep, 76
Volume, 60, 78, 81, 101
Volumetric, 20, 59
Vorotilov, 98, 100
Vortices, 95
Vossmeyer, T., 29, 53
VRH, 23
VT-STM, 35
V-V, 102, 105

W

W, 1–3, 9–11, 66, 107
Waals, 7, 8, 72, 81, 117
Waals-bonded, 8, 46, 62, 71
Wachter, P., 50
Wafers, 10, 12, 104
Wahyuni, S., 5
Walia, S., 25, 29
Wali, M., 5
Walls, 7, 34, 91, 93
Wang, B., 20, 31, 54, 85, 87
Wang, C., 14, 26, 27, 29, 30, 32, 52, 54, 68, 88, 97
Wang, D.-W., 31, 51, 52, 83, 87, 88
Wang, E., 28
Wang, F. F., 27, 51, 67, 98, 109, 115
Wang, G. X., 30, 31, 87, 88
Wang, H. G., 31, 32, 51, 52, 68, 83, 86–88
Wang, J., 26, 28, 29, 31, 51, 67, 83, 85–88, 97–100
Wang, J.-C., 70
Wang, J. H., 98
Wang, J. L., 84
Wang, J. W., 84
Wang, J. Y., 97
Wang, J.-Z., 87
Wang, K. F., 69, 88, 89, 97, 114
Wang, K. L., 84
Wang, K. X., 88, 114
Wang, L., 26, 28, 29, 30, 52, 53, 87–89
Wang, M. D., 28, 30, 53, 69, 84, 87
Wang, N., 54, 114, 115
Wang, P., 29, 52, 88, 97, 100
Wang, Q., 26, 29, 52, 84, 100
Wang, R. S., 27, 83, 88
Wang, S., 33, 67, 85–87, 113, 114
Wang, S. T., 28
Wang, S. Y., 115
Wang, T., 51, 86
Wang, W. Z., 28, 52, 54, 69, 98
Wang, X., 26, 28, 29, 31, 53, 67, 69, 85, 86, 89, 99, 100, 112
Wang, X. D., 84
Wang, X. G., 68
Wang, X. H., 98
Wang, X. J., 12, 29, 54
Wang, X. L., 51, 53, 87
Wang, Y., 29, 31, 32, 50–54, 69, 83, 86–88, 97, 100, 113
Wang, Y. F., 28
Wang, Y. Q., 67
Wang, Y. T., 16, 30
Wang, Y. Z., 100
Wang, Z. L., 5, 28, 29, 30, 32, 52, 54, 68, 69, 87, 88, 100, 114
Wan, J., 68
Wan, L., 89
Wan, S., 31, 51
Waqas, H., 5
Ward, T. Z., 99
Warwick, M., 113
Waser, R., 84
Wastewater, 43, 44, 55
Watanabe, M., 26
Watanabe, T., 33
Water, 1, 4, 7, 9, 15, 43, 44, 60, 67, 77–80, 110
water-soluble, 95, 104
water-splitting, 59
Wave, 2, 22, 34, 35, 38, 39, 50, 105–107, 111, 117, 118
Wavelengths, 23, 35, 38, 39
Weak\weakly, 8, 13, 15, 76
Wearable, 40
Weber, D. A., 27
Wehrspohn, R. B., 98, 100
Wei, B., 86
Wei, D.-H., 31

Wei, J., 98
Wei, M., 67
Wei, P., 52
Wei, R., 31
Wei, S., 31, 53, 54
Weitz, D. A., 114
Wei, W., 88
Wei, X. Z., 88, 98, 99
Wei, Y., 97, 115
Wei, Z., 29, 87
Wells, 8, 93
Wen, B., 86
Wen, C., 86
Wendorff, J. H., 98, 100
Wen, L., 31, 51
Wen, M., 13, 28
Weng, Q., 31
Weng, W. J., 97
Weng, Z., 86
Wen, Q.-Y., 114
Wenqing, Y., 114
Wen, R. T., 27
Wen, T.-L., 114
Wen, Y., 51, 86
Wen, Z., 69, 84, 87
Wenzel, S., 27
West, A. D., 84
Whang, W.-T., 30
White, 92
Whittingham, M. S., 28
Widiarti, N., 5
Wiley, 6, 9
Wills, J. M., 26
Windl, W., 25
Wire, 43
WIREs, 9
Wittmann, A., 115
Wlodarski, W., 53
WO_3, 4, 7–12, 15, 21, 23, 43, 44, 49, 50, 58, 59, 62, 65–67, 82, 117, 118
Wodtke, A. M., 53
Wolf, E., 112
Wongcharoen, N., 26
Wong, K., 84
Wong, S. S., 97–99
Woodward, P. M., 26
Woo, S., 84
W–O–W, 10, 11
Wrapping\wrapped, 43, 77, 80
Wrinkle-free, 74
Wu, C.-H., 26, 33, 68, 69, 86
Wu, D., 28, 86, 89, 97
Wu, G. J., 27, 33, 84, 88
Wu, H. B., 31, 85, 86, 97, 100
Wu, I.-H., 21, 31
Wu, J. B., 28, 29, 52, 54, 70, 85–87, 114, 115
Wu, K., 33
Wu, L., 32, 87
Wu, M., 30
Wu, N. Q., 99
Wu, P., 26
Wu, Q., 29–31, 116
Wuttig, M., 28
Wu, X.-L., 28, 51, 85, 86, 88
Wu, Y., 28, 32, 52, 68, 69
Wu, Y. Q., 116
Wu, Y.-X., 86
Wu, Z.-S., 29, 51, 52, 86

X

Xe, 110
Xenon, 61
Xia, B. Y., 85
Xiangfeng, C., 69
Xiang, H., 97
Xiang, J. Y., 67, 87
Xiang, Y., 33
Xiao, B., 30, 69
Xiao, F.-S., 87
Xiao, G., 100
Xiao, H., 32
Xiao, L., 86, 108, 115
Xiao, M., 69, 85
Xiao, X., 29, 69
Xiao, Y., 10, 27, 84, 87
Xiao, Z., 88
Xia, R., 67
Xia, X. H., 31, 51
Xi, C., 68
Xie, C., 30, 70, 85
Xie, D., 85
Xie, E., 53
Xie, J., 27, 28, 67, 85
Xie, K., 29
Xie, S. S., 30, 98
Xie, T., 32, 67
Xie, X. M., 88
Xie, Y., 28, 31, 86, 88
Xie, Y. K., 115
Xie, Z., 32
Xing, K., 29
Xing, L. L., 53, 86
Xing, R., 54
Xing, W., 89
Xing, X., 69, 86
Xingxing, G., 114
Xing, Y., 115
Xin, Y., 68, 83
Xiong, J., 29, 88

Xiong, Q., 86
Xiong, R., 13, 28
Xiong, S., 29
Xiong, W., 29
Xiong, X., 32
Xiong, Y., 114
Xiong, Z., 88
Xi, W., 83, 89
Xi, X. X., 84
X-ray, 37, 110
Xu, B., 51, 83, 88
Xu, C. Y., 69, 85, 86, 100
Xu, D., 99
Xue, C., 27, 87
Xue, D. S., 98
Xue, J., 32
Xue, S., 68
Xue, X. Y., 53, 86
Xue, Y., 99
Xu, G., 84, 97
Xu, H., 27, 32, 86
Xu, J. D., 67, 68, 97, 99
Xu, J. L., 85
Xu, K., 70, 83
Xu, L., 29, 54, 85, 86, 114
Xun, S., 30
Xu, Q., 30, 52
Xu, R. J., 85
Xu, S. Y., 68, 98, 115
Xu, W. C., 54, 67, 87, 115
Xu, X. Q., 28, 51, 52, 67, 85, 89, 98
Xu, Y. M., 28–31, 50, 51, 86, 88, 98
Xu, Z., 68, 69
Xylene, 15, 65

Y

Yagmur, V., 99
Yamada, H., 26
Yamagishi, A., 26
Yanagisawa, K., 97
Yan, C., 28, 68
YanChao, Z., 86
Yandong, B., 30
Yan, F., 85
Yang, B., 98
Yang, C. S., 25, 88
Yang, D., 30, 97, 99, 113
Yang, F. Y., 5, 100, 114
Yang, G., 32, 68
Yang, H. X., 29, 51, 88, 116
Yang, J., 28, 51, 69, 88
Yang, K. R., 29, 88
Yang, L., 69, 88
Yang, M., 26, 51
Yang, N., 26
Yang, P., 29
Yang, T., 12, 19, 30
Yang, Q.-H., 20, 31, 68, 87, 97, 99, 114
Yang, S. A., 26, 87, 98
Yang, S. M., 98
Yang, T., 28, 87
Yang, W. S., 25, 29, 30, 51, 54, 68, 85
Yang, W. Y., 98
Yang, X., 29, 32, 67, 86
Yang, Y., 26, 31, 52, 53, 86, 88, 97, 115, 116
Yang, Z. B., 27, 29, 52, 53, 69, 84, 85, 87, 89, 100
Yan, H. J., 84, 86
Yan, J. M., 28, 31, 51, 88
Yan, M., 51, 85
Yan, N., 86
Yan, Q. Y., 30, 68, 85, 86, 88
Yan, S., 85, 100
Yan, X., 28, 83, 87, 88
Yan, Y., 27
Yao, H. C., 28, 70
Yao, J., 27, 100
Yao, L., 54, 67
Yao, M., 31
Yao, Q., 88
Yao, T., 31
Yao, W., 27
Yao, X., 87
Yao, Y., 28, 86, 87
Yao, Z., 27
Yasuda, A., 29, 53
Yavuz, C. T., 53
Yazami, R. 31
Yazdani, A., 50
$YBa_2Cu_3O_7$, 2
Yeats, A. L., 83
Yee, K.-J., 112
Ye, H., 88
Yeh, P.-H., 31
Ye, K., 29
Yellow, 38
Yen, Y. C., 33
Yeo, J. Y., 51
Ye, T., 33
Yeung, C. W., 114
Ye, Z.-G., 26, 69, 100
Yi, C., 86
Yi, K., 68, 88
Yildiz, A., 23, 32
Yi, M., 26, 50
Yim, J. W. L., 115
Yin, C. Y., 69, 113
Yin, D., 87
Ying, G., 89
Yin, L.-C., 87
Yin, Y. B., 31
Yin, Z. Y., 32, 83

Yip, A. C. K., 67
Yi, R., 88
Yi, X., 26, 29, 104, 113, 115
Yi, Y., 26, 27
Yi, Z., 27, 85
Yolk-shell, 65
Yoon, C. M., 68
Yoon, D. H., 85, 98
Yoon, H., 108, 115
Yoon, J., 26, 31
Yoon, S. J., 27, 97
Yoon, Y. S., 27
You, J., 97
You, H., 113
You, K. H., 85
You, L., 83
Youn, D.-H., 113
Youqing, W., 30
You, X., 52
Yu, A., 83
Yuan, C., 88
Yuan, F., 28, 31
Yuan, S. G., 31, 51, 85
Yuan, S. L., 113
Yuan, Y., 97
Yuan, Z., 54
Yu, B. Y., 53, 98
Yu, C., 28, 32, 68
Yu, D., 69
Yu, G., 86
Yu, H., 29–31, 68, 69, 84, 97
Yu, J. C., 32, 53, 54, 67, 86, 89, 98, 99
Yu, J. Y., 88
Yu, K., 32, 88
Yu, L., 31, 88
Yun, K. Y., 30, 99
Yun, S. J., 112, 115
Yun, W. S., 97, 99
Yun, Y. J., 84, 98
Yu, P. F., 68
Yu, R. C., 25, 51
Yu, S. H., 86, 115
Yus, J., 69
Yusuf, M. H., 85
Yusufu, E., 69
Yuvaraja, S., 52
Yu, W.-R., 52
Yu, X., 52, 86
Yu, Y. T., 67, 68

Z

Zaanen, J., 50
Zacharias, M., 98
Zad, A. I., 27
Zafer, C., 5
Zaheb, H., 5
Zakharova, G. S., 30
Zakharov, N. D., 98, 100
Zaleska, A., 32
Zamora, F., 26
Zang, L., 32
Zang, Y. P., 85, 97, 100
Zappa, D., 32, 83
Zavabeti, A., 25, 29
Zeman, M., 113
Zener, C., 6
Zeng, C., 84, 98, 99
Zeng, D., 29, 70
Zeng, Q., 83
Zeng, W., 29
Zeng, Y. J., 26, 69, 85
Zeng, Z., 86
Zen, J.-M., 87
Zero, 1, 93
ZFO, 65
Zhai, B., 52
Zhai, H. Y., 99
Zhai, T., 25, 51
Zhai, X., 98
Zhang, B., 68, 69
Zhang, C. C., 26, 28, 31, 53, 68, 85, 88, 89, 97, 100
Zhang, C. J., 88
Zhang D. H., 86, 87, 98
Zhang D. L., 97
Zhang F., 32, 86, 87
Zhang G. Q., 26, 52, 68, 83, 85, 98
Zhang G. Y., 99
Zhang, H. -W., 25–27, 31, 52–54, 83, 85–88, 97, 99, 114
Zhang, J. Y., 5, 27–29, 32, 52–54, 70, 85, 86, 112
Zhang, K., 29, 54, 89
Zhang, L. H., 27, 32, 51, 53, 67, 70, 86, 114
Zhang, L. S., 88
Zhang, M., 27, 30, 32, 53, 54, 84, 86, 87, 100
Zhang, N., 69, 100
Zhang, P., 53, 67, 83, 87, 88, 114
Zhang, Q. F., 28, 29, 52, 54, 68, 83, 88, 115
Zhang, R. M., 27, 32, 68, 105, 113
Zhang, S. R., 26, 29, 54, 69, 84, 87, 88, 100, 115
Zhang, T. J., 28, 30, 52, 69, 87, 88, 97, 98, 113
Zhang, T. W., 100
Zhang, W. B., 22, 31, 32, 52, 67, 68, 86, 88, 115, 116
Zhang, W. J., 97
Zhang, X. B., 20, 27, 29–32, 51, 54, 67, 69, 85, 86, 88, 99, 114
Zhang, X. -F., 29
Zhang, X. G., 99
Zhang X. W., 85
Zhang X.-X., 86

Zhang X. Y., 97, 98
Zhang, Y. G., 5, 25, 28, 31, 32, 51, 53, 68, 69, 83, 85, 88, 97
Zhang, Y. H., 97
Zhang, Y. Q., 51
Zhang, Y. X., 28
Zhang, Z. B., 6, 26, 28, 52, 53, 67, 69, 86, 114
Zhang, Z. D., 28
Zhang, Z. H., 97
Zhan, R., 69
Zhan, T., 26
Zhao, B., 87
Zhao, C., 53, 88
Zhao, D.-L., 51, 116
Zhao, F., 27
Zhao, G., 52–54, 69
Zhao, H. S., 29, 30, 51, 68, 115
Zhaohui, L., 114
Zhao, J., 68, 88
Zhao, K., 51
Zhao, L. L., 98
Zhao, M. Q., 88, 116
Zhao, P. X., 84
Zhao, Q. D., 97, 115
Zhao, R., 54, 83
Zhao, T., 99
Zhao, W. F., 27, 115
Zhao, X. -B., 5, 28, 51, 52, 69, 98, 114
Zhao, X. J., 113
Zhao, X. S., 31
Zhao, X. Z., 98
Zhao, Y. F., 51, 54, 83, 85, 87, 88, 99, 100, 113
Zhao, Z. W., 67, 68
Zheng, C., 52
Zheng, F., 27, 29, 68
Zheng, H., 26, 115
Zheng, J. H., 54, 68, 86
Zheng, J. Y., 115
Zheng, L., 13, 28, 32
Zheng, S., 83
Zheng, W., 52, 54, 88, 99
Zheng, X., 29
Zheng, Y., 25
Zheng, Z., 28, 88
Zhen, Q., 68
Zhen, L., 100
Zhigalina, O. M., 98
Zhi, L., 31
Zhong, K., 29
Zhong, H. S., 31, 113
Zhong, Y., 53, 88
Zhou, A., 88
Zhou, C. W., 98
Zhou, D., 29, 31
Zhou, F., 100
Zhou, G., 86, 87
Zhou, H., 51, 83
Zhou, J., 27, 51, 53, 69, 83, 86, 88, 115
Zhou, L., 29, 88
Zhou, P., 88
Zhou, Q., 29, 67, 86
Zhou, S. H., 30, 67, 97
Zhou, T., 88
Zhou, W. Y., 32, 87, 98
Zhou, X. C., 29, 68, 99
Zhou, Y. L., 26, 27, 84, 85, 89, 98, 100, 107, 114
Zhou, Z., 27, 52, 100
Zhuang, H., 87
Zhuang, J., 53
Zhuang, Q.-C., 28, 86
Zhu, B., 70
Zhu, C., 83, 85, 86
Zhu, D. L., 67, 97
Zhu, G., 89
Zhu, H. W., 85, 97
Zhuiykov, S., 10, 27
Zhu, J. M., 69, 84–87, 89, 97, 100
Zhu, K., 31, 51
Zhu, L., 69
Zhuravlev, M. Y., 83, 84
Zhuo, S., 88
Zhu, Q.-C., 30, 86
Zhu, Q.-Z., 83, 88
Zhu, S., 67
Zhu, W., 26, 30
Zhu, X., 67, 87, 97
Zhu, X. D., 88
Zhu, X. H., 97
Zhu, Y., 5, 26, 28, 86
Zhu, Y. H., 31
Zhu, Y. L., 100
Zhu, Z., 85, 88
Ziebeck, K. R. A., 116
Ziese, M., 5, 99
Zigzag, 14
Zito, C. A., 27
Zlámal, M., 27
Zn, 111
ZnCo, 63
ZnFe, 65
ZnO, 23, 42, 48, 49, 58, 63, 66, 93, 107, 118
ZnTe, 93
Zohar, S. 116
Zone, 36
Zong, C., 86
Zou, K., 84
Zou, T., 115, 116
Zou, Y., 29, 30
Zou, Z., 32, 85
Zr, 1, 72, 73, 76, 90, 94

Zubko, P., 83, 100
Zuli, L., 30
Zylbersztejn, A., 103, 112

SYMBOLS

α, 12, 49
α-Fe, 44, 45
α-MoO, 12–15, 49
α-phase, 49
α-VO, 81
α-WO, 9
β, 12, 49
β-MoO, 12, 49
β-WO, 9
Γ, 16
γ, 16, 45
γ-Fe, 44, 45, 78
Δ, 38
δ, 35, 117
δ-functional, 106
ε, 36
ε-MnO, 82
Λ, 38
λ, 61
μA, 43
μm, 43, 91, 92, 109
ν, 104
Π, 17, 35, 39
π, 35, 102, 103
Σ, 3
σ, 3

For Product Safety Concerns and Information please contact our EU representative GPSR@taylorandfrancis.com
Taylor & Francis Verlag GmbH, Kaufingerstraße 24, 80331 München, Germany

www.ingramcontent.com/pod-product-compliance
Ingram Content Group UK Ltd.
Pitfield, Milton Keynes, MK11 3LW, UK
UKHW022124190926
475702UK00013B/1638
9781041034018